Current Trends in Symmetric Polynomials with Their Applications II

Current Trends in Symmetric Polynomials with Their Applications II

Editor

Taekyun Kim

MDPI • Basel • Beijing • Wuhan • Barcelona • Belgrade • Manchester • Tokyo • Cluj • Tianjin

Editor
Taekyun Kim
Department of Mathematics,
Kwangwoon University
Korea

Editorial Office
MDPI
St. Alban-Anlage 66
4052 Basel, Switzerland

This is a reprint of articles from the Special Issue published online in the open access journal *Symmetry* (ISSN 2073-8994) (available at: https://www.mdpi.com/journal/symmetry/special_issues/Symmetric_Polynomials).

For citation purposes, cite each article independently as indicated on the article page online and as indicated below:

LastName, A.A.; LastName, B.B.; LastName, C.C. Article Title. *Journal Name* **Year**, *Volume Number*, Page Range.

ISBN 978-3-0365-0360-8 (Hbk)
ISBN 978-3-0365-0361-5 (PDF)

Contents

About the Editor

Taekyun Kim received a PhD in the Department of Mathematics, Kyushu University in Japan(1994). He worked as a lecturer in Kyungpook National University in 1994–1996, a research professor in the Institute of Science Education, Kongju National University in 2001–2006, a professor(BK) in the Department of Electrical and Computer Engineering, Kyungpook National University in 2006–2008, and a chair professor in Tianjin Polytechnic University in 2015–2019. He has been working as a professor at the Department of Mathematics in Kwangwoon University since 2008 and has also been serving as the editor-in -chief in Advanced Studies in Contemporary Mathematics (http://www.jangjeonopen.or.kr/) since 1999.

Preface to "Current Trends in Symmetric Polynomials with Their Applications II"

The special numbers and polynomials play an extremely important role in various applications in such diverse areas as mathemaics, probability and statistics, mathematical physics, and engineering. Due to their powerful expressions, the combinations of special numbers and polynomials can be seen almost ubiquitously as the solutions of differential equations in the diverse fields by orthogonality condition, generating functions, recurrence relations, bosonic and fermionic p-adic integrals and etc.

Further, their importance can be also found in the developments of classical analysis, number theory, mathematical analysis, mathematical physics, symmetric functions, combinatorics, and other parts of the natural sciences.

In many years, a great amount of effort has been paid by many researchers to find new representations of families of special functions and polynomials with its practical applications.

This special issue will be contributed to the fields of special functions and orthogonal polynomials (or q-special functions and orthogonal polynomials) along the modern trends.

Taekyun Kim
Editor

Article

Some Identities on the Poly-Genocchi Polynomials and Numbers

Dmitry V. Dolgy [1] and **Lee-Chae Jang** [2,*]

[1] Kwangwoon Glocal Education Center, Kwangwoon University, Seoul 139-701, Korea; d_dol@mail.ru
[2] Graduate School of Education, Konkuk University, Seoul 143-701, Korea
* Correspondence: Lcjang@konkuk.ac.kr

Received: 28 May 2020; Accepted: 9 June 2020; Published: 14 June 2020

Abstract: Recently, Kim-Kim (2019) introduced polyexponential and unipoly functions. By using these functions, they defined type 2 poly-Bernoulli and type 2 unipoly-Bernoulli polynomials and obtained some interesting properties of them. Motivated by the latter, in this paper, we construct the poly-Genocchi polynomials and derive various properties of them. Furthermore, we define unipoly Genocchi polynomials attached to an arithmetic function and investigate some identities of them.

Keywords: polylogarithm functions; poly-Genocchi polynomials; unipoly functions; unipoly Genocchi polynomials

MSC: 11B83; 11S80

---

1. Introduction

The study of the generalized versions of Bernoulli and Euler polynomials and numbers was carried out in [1,2]. In recent years, various special polynomials and numbers regained the interest of mathematicians and quite a few results have been discovered. They include the Stirling numbers of the first and the second kind, central factorial numbers of the second kind, Bernoulli numbers of the second kind, Bernstein polynomials, Bell numbers and polynomials, central Bell numbers and polynomials, degenerate complete Bell polynomials and numbers, Cauchy numbers, and others (see [3–8] and the references therein). We mention that the study of a generalized version of the special polynomials and numbers can be done also for the transcendental functions like hypergeometric ones. For this, we let the reader refer to the papers [3,5,6,8,9]. The poly-Bernoulli numbers are defined by means of the polylogarithm functions and represent the usual Bernoulli numbers (more precisely, the values of Bernoulli polynomials at 1) when $k = 1$. At the same time, the degenerate poly-Bernoulli polynomials are defined by using the polyexponential functions (see [8]) and they are reduced to the degenerate Bernoulli polynomials if $k = 1$. The polyexponential functions were first studied by Hardy [10] and reconsidered by Kim [6,9,11,12] in view of an inverse to the polylogarithm functions which were studied by Zagier [13], Lewin [14], and Jaonquière [15]. In 1997, Kaneko [16] introduced poly-Bernoulli numbers which are defined by the polylogaritm function.

Recently, Kim-Kim introduced polyexponential and unipoly functions [9]. By using these functions, they defined type 2 poly-Bernoulli and type 2 unipoly-Bernoulli polynomials and obtained several interesting properties of them.

In this paper, we consider poly-Genocchi polynomials which are derived from polyexponential functions. Similarly motivated, in the final section, we define unipoly Genocchi polynomials attached to an arithmetic function and investigate some identities for them. In addition, we give explicit expressions and identities involving those polynomials.

It is well known, the Bernoulli polynomials of order α are defined by their generating function as follows (see [1–3,17,18]):

$$\left(\frac{t}{e^t - 1}\right)^{\alpha} e^{xt} = \sum_{n=0}^{\infty} B_n^{(\alpha)}(x)\frac{t^n}{n!}, \qquad . \tag{1}$$

We note that for $\alpha = 1$, $B_n(x) = B_n^{(1)}(x)$ are the ordinary Bernoulli polynomials. When $x = 0$, $B_n^{\alpha} = B_n^{\alpha}(0)$ are called the Bernoulli numbers of order α. The Genocchi polynomials $G_n(x)$ are defined by (see [19–24]).

$$\frac{2t}{e^t + 1}e^{xt} = \sum_{n=0}^{\infty} G_n(x)\frac{t^n}{n!}, \tag{2}$$

When $x = 0$, $G_n = G_n(0)$ are called the Genocchi numbers.

As is well-known, the Euler polynomials are defined by the generating function to be (see [1,4]).

$$\frac{2}{e^t + 1}e^{xt} = \sum_{n=0}^{\infty} E_n(x)\frac{t^n}{n!}, \tag{3}$$

For $n \geq 0$, the Stirling numbers of the first kind are defined by (see [5,7,25]),

$$(x)_n = \sum_{l=0}^{n} S_1(n,l)x^l, \tag{4}$$

where $(x)_0 = 1$, $(x)_n = x(x-1)\dots(x-n+1)$, $(n \geq 1)$. From (4), it is easy to see that

$$\frac{1}{k!}(\log(1+t))^k = \sum_{n=k}^{\infty} S_1(n,k)\frac{t^n}{n!}. \tag{5}$$

In the inverse expression to (4), for $n \geq 0$, the Stirling numbers of the second kind are defined by

$$x^n = \sum_{l=0}^{n} S_2(n,l)(x)_l. \tag{6}$$

From (6), it is easy to see that

$$\frac{1}{k!}(e^t - 1)^k = \sum_{n=k}^{\infty} S_2(n,k)\frac{t^n}{n!}. \tag{7}$$

2. The Poly-Genocchi Polynomials

For $k \in \mathbb{Z}$, by (2) and (14), we define the poly-Genocchi polynomials which are given by

$$\frac{2e_k(\log(1+t))}{e^t + 1}e^{xt} = \sum_{n=0}^{\infty} G_n^{(k)}(x)\frac{t^n}{n!}. \tag{8}$$

When $x = 0$, $G_n^{(k)} = G_n^{(k)}(0)$ are called the poly-Genocchi numbers. From (8), we see that

$$G_n^{(1)}(x) = G_n(x), \quad (n \in \mathbb{N} \cup \{0\}) \tag{9}$$

are the ordinary Genocchi polynomials. From (2), (4) and (8) , we observe that

$$\sum_{n=0}^{\infty} G_n^{(k)} \frac{t^n}{n!}$$

$$= \frac{2e_k(\log(1+t))}{e^t+1}$$

$$= \frac{2}{e^t+1} \sum_{m=1}^{\infty} \frac{(\log(1+t))^m}{(m-1)!m^k}$$

$$= \frac{2}{e^t+1} \sum_{m=0}^{\infty} \frac{(\log(1+t))^{m+1}}{m!(m+1)^k}$$

$$= \frac{2}{e^t-1} \sum_{m=0}^{\infty} \frac{1}{(m+1)^{k-1}} \sum_{l=m+1}^{\infty} S_1(l,m+1)\frac{t^l}{l!}$$

$$= \frac{2t}{e^t+1} \sum_{m=0}^{\infty} \frac{1}{(m+1)^{k-1}} \sum_{l=m}^{\infty} \frac{S_1(l+1,m+1)}{l+1} \frac{t^l}{l!}$$

$$= \left(\sum_{j=0}^{\infty} G_j \frac{t^j}{j!} \right) \sum_{l=0}^{\infty} \left(\sum_{m=0}^{l} \frac{1}{(m+1)^{k-1}} \frac{S_1(l+1,m+1)}{l+1} \right) \frac{t^l}{l!}$$

$$= \sum_{n=0}^{\infty} \left(\sum_{l=0}^{n} \sum_{m=0}^{l} \binom{n}{l} \frac{1}{(m+1)^{k-1}} \frac{S_1(l+1,m+1)}{l+1} G_{n-l} \right) \frac{t^n}{n!}. \tag{10}$$

Therefore, by (10), we obtain the following theorem.

Theorem 1. *For $k \in \mathbb{Z}$ and $n \in \mathbb{N} \cup \{0\}$, we have*

$$G_n^{(k)} = \sum_{l=0}^{n} \sum_{m=0}^{l} \binom{n}{l} \frac{1}{(m+1)^{k-1}} \frac{S_1(l+1,m+1)}{l+1} G_{n-l}. \tag{11}$$

Corollary 1. *For $n \in \mathbb{N} \cup \{0\}$, we have*

$$G_n^{(1)} = G_n = \sum_{l=0}^{n} \sum_{m=0}^{l} \binom{n}{l} \frac{S_1(l+1,m+1)}{l+1} G_{n-l}. \tag{12}$$

Moreover,

$$\sum_{l=1}^{n} \sum_{m=0}^{l} \binom{n}{l} \frac{S_1(l+1,m+1)}{l+1} G_{n-l} = 0, \quad (n \in \mathbb{N}). \tag{13}$$

Kim-Kim ([9]) defined the polyexponential function by (see [6,9–12,26]).

$$e_k(x) = \sum_{n=1}^{\infty} \frac{x^n}{(n-1)!n^k}, \tag{14}$$

In [18], it is well known that for $k \geq 2$,

$$\frac{d}{dx}e_k(x) = \frac{1}{x}e_{k-1}(x). \tag{15}$$

Thus, by (15), for $k \geq 2$, we get

$$e_k(x) = \int_0^x \frac{1}{t_1} \underbrace{\int_0^{t_1} \frac{1}{t_1} \cdots \int_0^{t_{k-2}} \frac{1}{t_{k-1}}}_{(k-2)\,times} (e^{t_{k-1}}-1) d_{k-1}t dt_{k-1} \cdots dt_1. \tag{16}$$

From (16), we obtain the following equation.

$$\sum_{n=0}^{\infty} G_n^{(k)} \frac{x^n}{n!}$$

$$= \frac{2}{e^x + 1} e_k(\log(1+x))$$

$$= \frac{2}{e^x + 1} \int_0^x \frac{1}{(1+t)\log(1+t)} e_{k-1}(\log(1+t))dt$$

$$= \frac{2}{e^x + 1} \int_0^x \frac{1}{(1+t_1)\log(1+t_1)}$$

$$\underbrace{\int_0^{t_1} \frac{1}{(1+t_2)\log(1+t_2)} \cdots \int_0^{t_{k-2}} \frac{t_{k-1}}{(1+t_{k-1})\log(1+t_{k-1})} dt_{k-1}dt_{k-2}\cdots dt_1, (k \geq 2).}_{(k-2)times} \quad (17)$$

Let us take $k = 2$. Then, by (2) and (16), we get

$$\sum_{n=0}^{\infty} G_n^{(2)} \frac{x^n}{n!} = \frac{2}{e^x + 1} \int_0^x \frac{t}{(1+t)\log(1+t)} dt$$

$$= \frac{2}{e^x + 1} \sum_{l=0}^{\infty} \frac{B_l^{(l)}}{l!} \int_0^x t^l dt$$

$$= \frac{2}{e^x + 1} \sum_{l=0}^{\infty} \frac{B_l^{(l)}}{l+1} \frac{x^{l+1}}{l!}$$

$$= \frac{2x}{e^x + 1} \sum_{l=0}^{\infty} \frac{B_l^{(l)}}{l+1} \frac{x^l}{l!}$$

$$= \left(\sum_{m=0}^{\infty} G_m \frac{x^m}{m!} \right) \left(\sum_{l=0}^{\infty} \frac{B_l^{(l)}}{l+1} \frac{x^l}{l!} \right)$$

$$= \sum_{n=0}^{\infty} \left(\sum_{l=0}^{n} \binom{n}{l} \frac{B_l^{(l)}}{l+1} G_{n-l} \right) \frac{x^n}{n!}. \quad (18)$$

Therefore, by (18), we obtain the following theorem.

Theorem 2. *Let $n \in \mathbb{N} \cup \{0\}$, we have*

$$G_n^{(2)} = \sum_{l=0}^{n} \binom{n}{l} \frac{B_l^{(l)}}{l+1} G_{n-l}. \quad (19)$$

From (3) and (16), we also get

$$
\begin{aligned}
\sum_{n=0}^{\infty} G_n^{(2)} \frac{x^n}{n!} &= \frac{2}{e^x + 1} \int_0^x \frac{t}{(1+t)\log(1+t)} dt \\
&= \frac{2}{e^x + 1} \sum_{l=0}^{\infty} B_l^{(l)} \frac{x^{l+1}}{(l+1)!} \\
&= \frac{2}{e^x + 1} \sum_{l=1}^{\infty} B_{l-1}^{(l-1)} \frac{x^l}{l!} \\
&= \left(\sum_{m=0}^{\infty} E_m \frac{x^m}{m!} \right) \left(\sum_{l=1}^{\infty} B_{l-1}^{(l-1)} \frac{x^l}{l!} \right) \\
&= \sum_{n=1}^{\infty} \left(\sum_{l=1}^{n} \binom{n}{l} B_{l-1}^{(l-1)} E_{n-l} \right) \frac{x^n}{n!}.
\end{aligned}
\tag{20}
$$

Therefore, by (20), we obtain the following theorem.

Theorem 3. *Let $n \geq 1$, we have*

$$
G_n^{(2)} = \sum_{l=1}^{n} \binom{n}{l} B_{l-1}^{(l-1)} E_{n-l}.
\tag{21}
$$

From (8), we observe that

$$
\begin{aligned}
\sum_{n=0}^{\infty} G_n^{(k)}(x) \frac{t^n}{n!} &= \frac{2 e_k(\log(1+t))}{e^t + 1} e^{xt} \\
&= \left(\sum_{l=0}^{\infty} G_l^{(k)} \frac{t^l}{l!} \right) \left(\sum_{m=0}^{\infty} x^m \frac{t^m}{m!} \right) \\
&= \sum_{n=0}^{\infty} \left(\sum_{l=0}^{n} \binom{n}{l} G_l^{(k)} x^{n-l} \right) \frac{t^n}{n!} \\
&= \sum_{n=0}^{\infty} \left(\sum_{l=0}^{n} \binom{n}{l} G_{n-l}^{(k)} x^l \right) \frac{t^n}{n!}.
\end{aligned}
\tag{22}
$$

From (22), we obtain the following theorem.

Theorem 4. *Let $n \in \mathbb{N}$, we have*

$$
G_n^{(k)}(x) = \sum_{l=0}^{n} \binom{n}{l} G_{n-l}^{(k)} x^l.
\tag{23}
$$

From (23), we observe that

$$
\begin{aligned}
\frac{d}{dx}G_n^{(k)}(x) &= \sum_{l=1}^{n}\binom{n}{l}G_{n-l}^{(k)}lx^{l-1}\\
&= \sum_{l=0}^{n-1}\binom{n}{l+1}G_{n-l-1}^{(k)}(l+1)x^l\\
&= \sum_{l=0}^{n-1}\frac{n!}{(l+1)!(n-l-1)!}G_{n-1-l}^{(k)}(l+1)x^l\\
&= n\sum_{l=0}^{n-1}\frac{(n-1)!}{l!(n-1-l)!}G_{n-1-l}^{(k)}x^l\\
&= nG_{n-1}^{(k)}(x).
\end{aligned}
\tag{24}
$$

From (24), we obtain the following theorem.

Theorem 5. *Let $n \in \mathbb{N}\cup\{0\}$ and $k \in \mathbb{Z}$, we have*

$$
\frac{d}{dx}G_n^{(k)}(x) = nG_{n-1}^{(k)}(x).
\tag{25}
$$

3. The Unipoly Genocchi Polynomials and Numbers

Let p be any arithmetic function which is real or complex valued function defined on the set of positive integers $\mathbb{N}$. Then, Kim-Kim ([9]) defined the unipoly function attached to polynomials by

$$
u_k(x|p) = \sum_{n=1}^{\infty}\frac{p(n)x^n}{n^k}, \quad (k \in \mathbb{Z}).
\tag{26}
$$

It is well known that

$$
u_k(x|1) = \sum_{n=1}^{\infty}\frac{x^n}{n^k} = Li_k(x)
\tag{27}
$$

is the ordinary polylogarithm function, and for $k \geq 2$,

$$
\frac{d}{dx}u_k(x|p) = \frac{1}{x}u_{k-1}(x|p),
\tag{28}
$$

and

$$
u_k(x|p) = \int_0^x \frac{1}{t}\underbrace{\int_0^t \frac{1}{t}\cdots\int_0^t \frac{1}{t}}_{(k-2)\,times}u_1(t|p)dtdt\cdots dt
\tag{29}
$$

By using (26), we define the unipoly Genocchi polynomials as follows:

$$
\frac{2}{e^t+1}u_k(\log(1+t)|p)e^{xt} = \sum_{n=0}^{\infty}G_{n,p}^{(k)}(x)\frac{t^n}{n!}.
\tag{30}
$$

Let us take $p(n) = \frac{1}{(n-1)!}$. Then we have

$$
\begin{aligned}
\sum_{n=0}^{\infty} G_{n,p}^{(k)}(x) \frac{t^n}{n!} &= \frac{2}{e^t+1} u_k \left(\log(1+t) \,\middle|\, \frac{1}{(n-1)!} \right) e^{xt} \\
&= \frac{2}{e^t+1} \sum_{m=1}^{\infty} \frac{(\log(1+t))^m}{m^k (m-1)!} e^{xt} \\
&= \frac{2e_k(\log(1+t))}{e^t+1} e^{xt} \\
&= \sum_{n=0}^{\infty} G_n^{(k)}(x) \frac{t^n}{n!}.
\end{aligned}
\tag{31}
$$

Thus, by (31), we have the following theorem.

Theorem 6. *If we take* $p(n) = \frac{1}{(n-1)!}$ *for* $n \in \mathbb{N} \cup \{0\}$ *and* $k \in \mathbb{Z}$, *then we have*

$$
G_{n,p}^{(k)}(x) = G_n^{(k)}(x).
\tag{32}
$$

From (4) and (30) with $x = 0$, we have

$$
\begin{aligned}
&\sum_{n=0}^{\infty} G_{n,p}^{(k)} \frac{t^n}{n!} \\
&= \frac{2}{e^t+1} \sum_{m=1}^{\infty} \frac{p(m)}{m^k} (\log(1+t))^m \\
&= \frac{2}{e^t+1} \sum_{m=0}^{\infty} \frac{p(m+1)(m+1)!}{(m+1)^k} \sum_{l=m+1}^{\infty} S_1(l, m+1) \frac{t^l}{l!} \\
&= \frac{2}{e^t+1} \sum_{m=0}^{\infty} \frac{p(m+1)(m+1)!}{(m+1)^k} \sum_{l=m}^{\infty} S_1(l+1, m+1) \frac{t^{l+1}}{(l+1)!} \\
&= \frac{2t}{e^t+1} \sum_{m=0}^{\infty} \frac{p(m+1)(m+1)!}{(m+1)^k} \sum_{l=m}^{\infty} S_1(l+1, m+1) \frac{t^l}{(l+1)!} \\
&= \left(\sum_{j=0}^{\infty} G_j \frac{t^j}{j!} \right) \sum_{l=0}^{\infty} \left(\sum_{m=0}^{l} \frac{p(m+1)(m+1)!}{(m+1)^k} \frac{S_1(l+1, m+1)}{l+1} \right) \frac{t^l}{l!} \\
&= \sum_{n=0}^{\infty} \left(\sum_{l=0}^{n} \sum_{m=0}^{l} \binom{n}{l} \frac{p(m+1)(m+1)!}{(m+1)^k} \frac{S_1(l+1, m+1)}{l+1} G_{n-l} \right) \frac{t^n}{n!}.
\end{aligned}
\tag{33}
$$

Therefore, by comparing the coefficients on both sides of (33), we obtain the following theorem.

Remark 1. *Let* $n \in \mathbb{N}$ *and* $k \in \mathbb{Z}$. *Then, we have*

$$
G_{n,p}^{(k)} = \sum_{l=0}^{n} \sum_{m=0}^{l} \binom{n}{l} \frac{p(m+1)(m+1)!}{(m+1)^k} \frac{S_1(l+1, m+1)}{l+1} G_{n-l}.
\tag{34}
$$

In particular,

$$
G_{n, \frac{1}{(n-1)!}}^{(k)} = \sum_{l=0}^{n} \sum_{m=0}^{l} \binom{n}{l} \frac{G_{n-l}}{(m+1)^{k-1}} \frac{S_1(l+1, m+1)}{l+1}
\tag{35}
$$

arrives at (11).

From (30), we easily obtain the following theorem.

Theorem 7. *Let $n \in \mathbb{N} \cup \{0\}$ and $k \in \mathbb{Z}$. Then, we have*

$$G_{n,p}^{(k)}(x) = \sum_{l=0}^{n} \binom{n}{l} G_{n-l,p}^{(k)} x^l. \tag{36}$$

From (36), we easily obtain the following theorem.

Theorem 8. *Let $n \in \mathbb{N} \cup \{0\}$ and $k \in \mathbb{Z}$. Then, we have*

$$\frac{d}{dx} G_{n,p}^{(k)}(x) = n G_{n-1,p}^{(k)}(x). \tag{37}$$

Finally, by (4) and (30), we observe that

$$
\begin{aligned}
&\sum_{n=0}^{\infty} G_{n,p}^{(k)} \frac{t^n}{n!} \\
&= \frac{2}{e^t + 1} \sum_{m=1}^{\infty} \frac{p(m)}{m^k} \frac{m!}{m!} (\log(1+t))^m \\
&= \frac{2}{e^t + 1} \sum_{m=1}^{\infty} \frac{p(m+1)}{(m+1)^k} \frac{(m+1)!}{(m+1)!} (\log(1+t))^{m+1} \\
&= \sum_{j=0}^{\infty} E_j \frac{t^j}{j!} \sum_{m=0}^{\infty} \frac{p(m+1)(m+1)!}{(m+1)^k} \sum_{l=m+1}^{\infty} S_1(l, m+1) \frac{t^l}{l!} \\
&= \sum_{j=0}^{\infty} E_j \frac{t^j}{j!} \sum_{l=0}^{\infty} \sum_{m=0}^{l} \frac{p(m+1)(m+1)!}{(m+1)^k} \sum_{l=m}^{\infty} S_1(l+1, m+1) \frac{t^{l+1}}{(l+1)!} \\
&= \sum_{n=0}^{\infty} \left(\sum_{l=0}^{n} \sum_{m=0}^{l} \binom{n}{l} \frac{p(m+1)(m+1)!}{(m+1)^k} \frac{S_1(l+1, m+1)}{l+1} E_{n-l} \right) \frac{t^n}{n!}.
\end{aligned}
\tag{38}
$$

From (37), we obtain the following theorem.

Theorem 9. *Let $n \in \mathbb{N}$ and $k \in \mathbb{Z}$, we have*

$$G_{n,p}^{(k)} = \sum_{l=0}^{n} \sum_{m=0}^{l} \binom{n}{l} \frac{p(m+1)(m+1)!}{(m+1)^k} \frac{S_1(l+1, m+1)}{l+1} E_{n-l}. \tag{39}$$

4. Conclusions

In 2019, Kim-Kim considered the polyexponential functions and poly-Bernoulli polynomials. In the same view as these functions and polynomials, we defined the poly-Genocchi polynomials (Equation (8)) and obtained some identities (Theorem 1 and Corollary 1). In particular, we observed explicit poly-Genocchi numbers for $k = 2$ (Theorems 2, 3 and 4). Furthermore, by using the unipoly functions, we defined the unipoly Genocchi polynomials (Equation (30)) and obtained some their properties (Theorems 6 and 7). Finally, we obtained the derivative of the unipoly Genocchi polynomials (Theorem 8) and gave the identity indicating the relationship of unipoly Genocchi polynomials and Euler polynomials (Theorem 9). It is recommended that our readers look at references [27–31] if they want to know the applications related to this paper.

Author Contributions: L.-C.J. and D.V.D. conceived the framework and structured the whole paper; D.V.D. and L.-C.J. checked the results of the paper and completed the revision of the article. All authors have read and agreed to the published version of the manuscript.

Funding: The present research has been conducted by the Research Grant of Kwangwoon University in 2020.

Conflicts of Interest: The authors declare no conflict of interest.

References

1. Bayad, A.; Kim, T. Identities for the Bernoulli, the Euler and the Genocchi numbers and polynomials. *Adv. Stud. Contemp. Math. (Kyungshang)* **2010**, *20*, 247–253.
2. Bayad, A.; Chikhi, J. Non linear recurrences for Apostol-Bernoulli-Euler numbers of higher order. *Adv. Stud. Contemp. Math. (Kyungshang)* **2012**, *22*, 1–6.
3. Kim, T.; Kim, D.S.; Lee, H.; Kwon, J. Degenerate binomial coefficients and degenerate hypergeometric functions. *Adv. Differ. Equ.* **2020**, *2020*, 115. [CrossRef]
4. Kim, T. Some identities for the Bernoulli, the Euler and the Genocchi numbers and polynomials. *Adv. Stud. Contemp. Math. (Kyungshang)* **2010**, *20*, 23–28.
5. Kim, T.; Kim, D.S. Some identities of extended degenerate r-central Bell polynomials arising from umbral calculus. *Rev. R. Acad. Cienc. Exactas Fis. Nat. Ser. A Mat. RASAM* **2020**, *114*, 1. [CrossRef]
6. Kim, T.; Kim, D.S. Degenerate polyexponential functions and degenerate Bell polynomials. *J. Math. Anal. Appl.* **2020**, *487*, 124017. [CrossRef]
7. Kim, T.; Kim, D.S. A note on central Bell numbers and polynomilas. *Russ. J. Math. Phys.* **2020**, *27*, 76–81.
8. Kim, T.; Kim, D.S.; Kim, H.Y.; Jang, L.C. Degenerate poly-Bernoulli numbers and polynomials. *Informatica* **2020**, *31*, 1–7.
9. Kim, D.S.; Kim, T. A note on polyexponential and unipoly functions. *Russ. J. Math. Phys.* **2019**, *26*, 40–49. [CrossRef]
10. Hardy, G.H. On a class a functions. *Proc. Lond. Math. Soc.* **1905**, *3*, 441–460. [CrossRef]
11. Kim, T.; Kim, D.S.; Kwon, J.K.; Lee, H.S. Degenerate polyexponential functions and type 2 degenerate poly-Bernoulli numbers and polynomials. *Adv. Differ. Equ.* **2020**, *2020*, 168. [CrossRef]
12. Kim, T.; Kim, D.S.; Kwon, J.K.; Kim, H.Y. A note on degenerate Genocchi and poly-Genocchi numbers and polynomials. *J. Inequal. Appl.* **2020**, *2020*, 110. [CrossRef]
13. Zagier, D. The Bloch-Wigner-Ramakrishnan polylogarithm function. *Math. Ann.* **1990**, *286*, 613–624. [CrossRef]
14. Lewin, L. *Polylogarithms and Associated Functions*; With a foreword by A. J. Van der Poorten; North-Holland Publishing Co.: New York, NY, USA; Amsterdam, The Netherlands, 1981; p. xvii+359.
15. Jonquière, A. Note sur la serie $\sum_{m=1}^{\infty} \frac{x^n}{n^s}$. *Bull. Soc. Math. France* **1889**, *17*, 142–152.
16. Kaneko, M. Poly-Bernoulli numbers. *J. Theor. Nombres Bordeaux* **1997**, *9*, 221–228. [CrossRef]
17. Dolgy, D.V.; Kim, T.; Kwon, H.-I.; Seo, J.J. Symmetric identities of degenerate q-Bernoulli polynomials under symmetric group S_3. *Proc. Jangjeon Math. Soc.* **2016**, *19*, 1–9.
18. Gaboury, S.; Tremblay, R.; Fugere, B.-J. Some explicit formulas for certain new classes of Bernoulli, Euler and Genocchi polynomials. *Proc. Jangjeon Math. Soc.* **2014**, *17*, 115–123.
19. Cangul, I.N.; Kurt, V.; Ozden, H.; Simsek, Y. On the higher-order $w - q$-Genocchi numbers. *Adv. Stud. Contemp. Math. (Kyungshang)* **2009**, *19*, 39–57.
20. Duran, U.; Acikgoz, M.; Araci, S. Symmetric identities involving weighted q-Genocchi polynomials under S_4. *Proc. Jangjeon Math. Soc.* **2015**, *18*, 445–465.
21. Jang, L.C.; Ryoo, C.S.; Lee, J.G.; Kwon, H.I. On the k-th degeneration of the Genocchi polynomials. *J. Comput. Anal. Appl.* **2017**, *22*, 1343–1349.
22. Jang, L.C. A study on the distribution of twisted q-Genocchi polynomials. *Adv. Stud. Contemp. Math. (Kyungshang)* **2009**, *18*, 181–189.
23. Kim, D.S.; Kim, T. A note on a new type of degenerate Bernoulli numbers. *Russ. J. Math. Phys.* **2020**, *27*, 227–235. [CrossRef]
24. Kurt, B.; Simsek, Y. On the Hermite based Genocchi polynomials. *Adv. Stud. Contemp. Math. (Kyungshang)* **2013**, *23*, 13–17.
25. Kwon, J.; Kim, T.; Kim, D.S.; Kim, H.Y. Some identities for degenerate complete and inomplete r-Bell polynomials. *J. Inequal. Appl.* **2020**, *23*. [CrossRef]
26. Roman, S. *The Umbral Calculus, Pure and Applied Mathematics, 111*; Academic Press, Inc.: Harcourt Brace Jovanovich: New York, NY, USA, 1984; p. x+193.
27. Jang, L.-C.; Kim, D.S.; Kim, T.; Lee, H. p-Adic integral on $\mathbb{Z}_p$ associated with degenerate Bernoulli polynomials of the second kind. *Adv. Differ. Equ.* **2020**, *2020*, 278. [CrossRef]

28. kim, T.; Kim, D.S.; Jang, L.-C.; Lee, H. Jindalrae and Gaenari numbers and polynomials in connection with Jindalrae–Stirling numbers. *Adv. Differ. Equ.* **2020**, *2020*, 245. [CrossRef]
29. Kim, T.; Kim, D.S. Some relations of two type 2 polynomials and discrete harmonic numbers and polynomials. *Symmetry* **2020**, *12*, 905. [CrossRef]
30. Kwon, J.; Jang, L.-C. A note on the type 2 poly-Apostol-Bernoulli polynomials. *Adv. Stud. Contemp. Math. (Kyungshang)* **2020**, *30*, 253–262.
31. Kim, D.S.; Kim, T.; Kwon, J.; Lee, H. A note on λ-Bernoulli numbers of the second kind. *Adv. Stud. Contemp. Math. (Kyungshang)* **2020**, *30*, 187–195.

 *symmetry*

Article

An Erdős-Ko-Rado Type Theorem via the Polynomial Method

Kyung-Won Hwang [1,†], **Younjin Kim** [2,*,†] and **Naeem N. Sheikh** [3,†]

[1] Department of Mathematics, Dong-A University, Busan 49315, Korea; khwang@dau.ac.kr
[2] Institute of Mathematical Sciences, Ewha Womans University, Seoul 03760, Korea
[3] School of Sciences and Engineering, Al Akhawayn University, 53000 Ifrane, Morocco; n.sheikh@aui.ma
[*] Correspondence: younjinkim@ewha.ac.kr; Tel.: +82-2-3277-6993
[†] These authors contributed equally to this work.

Received: 29 March 2020; Accepted: 16 April 2020; Published: 17 April 2020

Abstract: A family $\mathcal{F}$ is an intersecting family if any two members have a nonempty intersection. Erdős, Ko, and Rado showed that $|\mathcal{F}| \leq \binom{n-1}{k-1}$ holds for a k-uniform intersecting family $\mathcal{F}$ of subsets of $[n]$. The Erdős-Ko-Rado theorem for non-uniform intersecting families of subsets of $[n]$ of size at most k can be easily proved by applying the above result to each uniform subfamily of a given family. It establishes that $|\mathcal{F}| \leq \binom{n-1}{k-1} + \binom{n-1}{k-2} + \cdots + \binom{n-1}{0}$ holds for non-uniform intersecting families of subsets of $[n]$ of size at most k. In this paper, we prove that the same upper bound of the Erdős-Ko-Rado Theorem for k-uniform intersecting families of subsets of $[n]$ holds also in the non-uniform family of subsets of $[n]$ of size at least k and at most $n - k$ with one more additional intersection condition. Our proof is based on the method of linearly independent polynomials.

Keywords: Erdős-Ko-Rado theorem; intersecting families; polynomial method

1. Introduction

Let $[n]$ be the set $\{1, 2, \cdots, n\}$. A family $\mathcal{F}$ of subsets of $[n]$ is *intersecting* if $F \cap F'$ is non-empty for all $F, F' \in \mathcal{F}$. A family $\mathcal{F}$ of subsets of $[n]$ is *t-intersecting* if $|F \cap F'| \geq t$ holds for any $F, F' \in \mathcal{F}$. A family $\mathcal{F}$ is *k-uniform* if it is a collection of k-subsets of $[n]$. In 1961, Erdős, Ko, and Rado [1] were interested in obtaining an upper bound on the maximum size that an intersecting k-uniform family can have and proved the following theorem which bounds the cardinality of an intersecting k-uniform family.

Theorem 1 (Erdős-Ko-Rado Theorem [1]). *If $n \geq 2k$ and $\mathcal{F}$ is an intersecting k-uniform family of subsets of $[n]$, then*

$$|\mathcal{F}| \leq \binom{n-1}{k-1}.$$

Erdős-Ko-Rado Theorem is an important result of extremal set theory and has been an inspiration for various generalizations by many authors for over 50 years. Erdős, Ko, and Rado [1] also proved that there exists an integer $n_0(k, t)$ such that if $n \geq n_0(k, t)$, then the maximum size of a t-intersecting k-uniform family of subsets of $[n]$ is $\binom{n-t}{k-t}$. The following generalization of the Erdős-Ko-Rado Theorem was proved by Frankl [2] for $t \geq 15$, and was completed by Wilson [3] for all t. It establishes that the generalized EKR theorem is true if $n \geq (k - t + 1)(t + 1)$.

Theorem 2 (Generalized Erdős-Ko-Rado Theorem [2,3]). *If $n \geq (k - t + 1)(t + 1)$ and $\mathcal{F}$ is a t-intersecting k-uniform family of subsets of $[n]$, then we have*

$$|\mathcal{F}| \leq \binom{n-t}{k-t}.$$

The Erdős-Ko-Rado Theorem can be restated as follows.

Theorem 3 (Erdős-Ko-Rado Theorem [1]). *If $\mathcal{F}$ is a family of subsets F_i of $[n]$ with $|F_i| = k$ and $|F_i| \leq n - k$ that satisfies the following two conditions, for $i \neq j$*

(a) $1 \leq |F_i \cap F_j| \leq k - 1$
(b) $1 \leq |F_i \cap F_j^c| \leq k - 1$

then we have

$$|\mathcal{F}| \leq \binom{n-1}{k-1}.$$

2. Results

The following EKR-type theorem for non-uniform intersecting families of subsets of $[n]$ of size at most k can be easily proved by applying Theorem 3 to each uniform subfamily of the given non-uniform family.

Theorem 4. *If $\mathcal{F}$ is a family of subsets F_i of $[n]$, with $|F_i| \leq k$ and $n \geq 2k$, that satisfies the following two conditions, for $i \neq j$*

(a) $1 \leq |F_i \cap F_j| \leq k - 1$
(b) $1 \leq |F_i \cap F_j^c| \leq k - 1$

then we have

$$|\mathcal{F}| \leq \binom{n-1}{k-1} + \binom{n-1}{k-2} + \cdots + \binom{n-1}{0}.$$

In 2014, Alon, Aydinian, and Huang [4] gave the following strengthening of the bounded rank Erdős-Ko-Rado theorem by obtaining the same upper bound under a weaker condition as follows.

Theorem 5 (Alon, Aydinian, and Huang [4]). *Let $\mathcal{F}$ be a family of subsets of $[n]$ of size at most k, $1 \leq k \leq n - 1$. Suppose that for every two subsets $A, B \in \mathcal{F}$, if $A \cap B = \varnothing$, then $|A| + |B| \leq k$. Then we have*

$$|\mathcal{F}| \leq \binom{n-1}{k-1} + \binom{n-1}{k-2} + \cdots + \binom{n-1}{0}.$$

Since the bound $\binom{n-1}{k-1} + \binom{n-1}{k-2} + \cdots + \binom{n-1}{0}$ is much larger than $\binom{n-1}{k-1}$, this leads to the following interesting question: when is it possible to get the same bound as in the Erdős-Ko-Rado theorem for uniform intersecting families for the non-uniform intersecting families? We answer this question in the main result of this paper, where we prove that the same upper bound of the EKR Theorem for k-uniform intersecting families of subsets of $[n]$ also holds in the non-uniform family of subsets of $[n]$ of size at least k and at most $n - k$ with one more additional intersection condition, as follows.

Theorem 6. *If $\mathcal{F}$ is a family of subsets F_i of $[n]$ with $k \leq |F_i| \leq n - k$ that satisfies the following three conditions, for $i \neq j$*

(a) $1 \leq |F_i \cap F_j| \leq k - 1$
(b) $1 \leq |F_i \cap F_j^c| \leq k - 1$
(c) $1 \leq |F_i^c \cap F_j^c| \leq k - 1$

then we have

$$|\mathcal{F}| \leq \binom{n-1}{k-1}.$$

Please note that if we remove the third condition in Theorem 6, we get the same bound of the Erdős-Ko-Rado theorem for k-uniform intersecting families under the same condition for subsets of $[n]$ that are of size at least k and at most $n - k$.

Erdős-Ko-Rado Theorem is a seminal result in extremal combinatorics and has been proved by various methods (see a survey in [5]). There have been many results that have generalized EKR in various ways over the decades. The aim of this paper is to give a generalization of the EKR Theorem to non-uniform families with some extra conditions. Our proof is based on the method of linearly independent multilinear polynomials.

Our paper is organized as follows. In Section 3, we will introduce our main tool, the method of linearly independent multilinear polynomials. In Section 4, we will give the proof of our main result, Theorem 6.

3. Polynomial Method

The method of linearly independent polynomials is one of the most powerful methods for counting the number of sets in various combinatorial settings. In this method, we correspond multilinear polynomials to the sets and then prove that these polynomials are linearly independent in some space. In 1975, Ray-Chaudhuri and Wilson [6] obtained the following result by using the method of linearly independent polynomials.

Theorem 7 (Ray-Chaudhuri and Wilson [6]). *Let $l_1, l_2, \cdots, l_s < n$ be nonnegative integers. If $\mathcal{F}$ is a k-uniform family of subsets of $[n]$ such that $|A \cap B| \in L = \{l_1, l_2, \cdots, l_s\}$ holds for every pair of distinct subsets $A, B \in \mathcal{F}$, then $|\mathcal{F}| \leq \binom{n}{s}$ holds.*

In 1981, Frankl and Wilson [7] obtained the following nonuniform version of the Ray-Chaudhuri-Wilson Theorem using the polynomial method. Their proof is given underneath.

Theorem 8 (Frankl and Wilson [7]). *Let $l_1, l_2, \cdots, l_s < n$ be nonnegative integers. If $\mathcal{F}$ is a family of subsets of $[n]$ such that $|A \cap B| \in L = \{l_1, l_2, \cdots, l_s\}$ holds for every pair of distinct subsets $A, B \in \mathcal{F}$, then $|\mathcal{F}| \leq \sum_{k=0}^{s} \binom{n}{k}$ holds.*

Proof. Let x be the n-tuple of variables $x_1, x_2, \cdots, x_n$, where x_i takes the values only 0 and 1. Then all the polynomials we will work with have the relation $x_i^2 = x_i$ in their domain. Let $F_1, F_2, \cdots, F_m$ be the distinct sets in $\mathcal{F}$, listed in non-decreasing order according to their sizes. We define the characteristic vector $v_i = (v_{i_1}, v_{i_2}, \cdots, v_{i_n})$ of F_i such that $v_{i_j} = 1$ if $j \in F_i$ and $v_{i_j} = 0$ if $j \notin F_i$. We consider the following multilinear polynomial

$$f_i(x) = \prod_{l \in L,\ l < |F_i|} (v_i \cdot x - l)$$

where $x = (x_1, x_2, \cdots, x_n)$.

Then we obtain that $f_i(v_i) \neq 0$ and $f_i(v_j) = 0$ for $j < i$. As the vectors v_i are $0 - 1$ vectors, we have an another multilinear polynomial $g_i(x)$ such that $f_i(x) = g_i(x)$ holds for all $x \in \{0, 1\}^n$ by substituting x_k for the powers of x_k, where $k = 1, 2, \cdots, n$. Then it is easy to see that the polynomials $g_1, g_2, \cdots, g_m$ are linearly independent over $\mathbb{R}$. Since the dimension of n-variable multilinear polynomials of degree at most s is $\sum_{k=0}^{s} \binom{n}{k}$, we have

$$|\mathcal{F}| \leq \sum_{k=0}^{s} \binom{n}{k}$$

finishing the proof of Theorem 8. $\square$

In the same paper, Frankl and Wilson [7] obtained the following modular version of Theorem 7.

Theorem 9 (Frankl and Wilson [7]). *If $\mathcal{F}$ is a family of subsets of $[n]$ such that $|A \cap B| \equiv l \in L \pmod{p}$ holds for every pair of distinct subsets $A, B \in \mathcal{F}$, then $|\mathcal{F}| \leq \binom{n}{|L|}$ holds.*

In 1983, Deza, Frankl and Singhi [8] obtained the following modular version of Theorem 8.

Theorem 10 (Deza, Frankl and Singhi [8]). *If $\mathcal{F}$ is a family of subsets of $[n]$ such that $|A \cap B| \equiv l \in L \pmod{p}$ holds for every pair of distinct subsets $A, B \in \mathcal{F}$ and $|A| \not\equiv l \pmod{p}$ for every $A \in \mathcal{F}$, then $|\mathcal{F}| \leq \sum_{i=0}^{|L|} \binom{n}{i}$ holds.*

In 1991, Alon, Babai, and Suzuki [9] gave another modular version of Theorem 8 by replacing the condition of nonuniformity with the condition that the members of $\mathcal{F}$ have r different sizes as follows. Their proof was also based on the polynomial method.

Theorem 11 (Alon-Babai-Suzuki [9]). *Let $K = \{k_1, k_2, \cdots, k_r\}$ and $L = \{l_1, l_2, \cdots, l_s\}$ be two disjoint subsets of $\{0, 1, \cdots, p-1\}$, where p is a prime, and let $\mathcal{F}$ be a family of subsets of $[n]$ whose sizes modulo p are in the set K, and $|A \cap B| \pmod{p} \in L$ holds for every distinct two subsets A, B in $\mathcal{F}$, then the largest size of such a family $\mathcal{F}$ is $\binom{n}{s} + \binom{n}{s-1} + \cdots + \binom{n}{s-r+1}$ under the conditions $r(s-r+1) \leq p-1$ and $n \geq s + \max_{1 \leq i \leq r} k_i$.*

In the same paper, Alon, Babai, and Suzuki [9] also conjectured that the statement of Theorem 11 remains true if the condition $r(s-r+1) \leq p-1$ is dropped. Recently Hwang and Kim [10] proved this conjecture of Alon, Babai and Suzuki (1991), using the method of linearly independent polynomials. This result is as follows.

Theorem 12 (Hwang and Kim [10]). *Let $K = \{k_1, k_2, \cdots, k_r\}$ and $L = \{l_1, l_2, \cdots, l_s\}$ be two disjoint subsets of $\{0, 1, \cdots, p-1\}$, where p is a prime, and let $\mathcal{F}$ be a family of subsets of $[n]$ whose sizes modulo p are in the set K, and $|A \cap B| \pmod{p} \in L$ for every distinct two subsets A, B in $\mathcal{F}$, then the largest size of such a family $\mathcal{F}$ is $\binom{n}{s} + \binom{n}{s-1} + \cdots + \binom{n}{s-r+1}$ under the only condition that $n \geq s + \max_{1 \leq i \leq r} k_i$.*

The method of linearly independent polynomials has also been used to prove many intersection theorems about set families by Blokhuis [11], Chen and Liu [12], Furedi, Hwang, and Weichsel [13], Liu and Yang [14], Qian and Ray-Chaudhuri [15], Ramanan [16], Snevily [17,18], Wang, Wei, and Ge [19], and others.

4. Proof of the Main Result

In this section, we prove Theorem 6. As we have mentioned before, our proof is based on the polynomial method. Let x be the n-tuple of variables $x_1, x_2, \cdots, x_n$, where x_i takes the values only 0 and 1. Then all the polynomials we will work with have the relation $x_i^2 = x_i$ in their domain.

Proof of Theorem 6. The result is immediate if $|\mathcal{F}| = 1$. Suppose $|\mathcal{F}| > 1$. Let $F_1, F_2, \cdots, F_f$ be the distinct sets in $\mathcal{F}$, listed in non-decreasing order of size. We define the characteristic vector $v_i = (v_{i_1}, v_{i_2}, \cdots, v_{i_n})$ of F_i such that $v_{i_j} = 1$ if $j \in F_i$ and $v_{i_j} = 0$ if $j \notin F_i$.

We consider the following family of multilinear polynomials

$$f_i(x) = \prod_{j=1}^{k-1} (v_i \cdot x - j)$$

where $x = (x_1, x_2, \cdots, x_n)$.

Since $|F_1| \leq |F_2|$, there exists some $p \in F_2$ such that $p \notin F_1$. Let $\mathcal{G} = \{G_1, G_2, \cdots, G_g\}$ be the family of subsets of $[n]$ with the size at most $k-2$, which is listed in non-decreasing order of size, and not containing p. Next, we consider the second family of multilinear polynomials

$$g_i(x) = (x_p - 1) \prod_{j \in G_i} x_j$$

where $1 \le i \le g$. Let $\mathcal{H} = \{H_1, H_2, \cdots, H_h\}$ be the family of subsets of $[n]$ with the size at most $k-1$, which is listed in non-decreasing order of size, and containing p. Then, we consider our third and last family of multilinear polynomials

$$h_i(x) = \prod_{j=0}^{|H_i|-1}(w_i \cdot x - j) - \sum_{l:p \notin F_l} \frac{\prod_{j=0}^{|H_i|-1}(w_i \cdot v_l^c - j)}{\prod_{j=1}^{k-1}(v_l^c \cdot v_l^c - j)} \prod_{j=1}^{k-1}(v_l^c \cdot x - j)$$
$$- \sum_{l:p \in F_l} \frac{\prod_{j=0}^{|H_i|-1}(w_i \cdot v_l - j)}{\prod_{j=1}^{k-1}(v_l \cdot v_l - j)} \prod_{j=1}^{k-1}(v_l \cdot x - j)$$

where w_i is the characteristic vector of H_i.

We claim that the functions $f_i(x)$, $g_i(x)$, and $h_i(x)$ taken together are linearly independent. Assume that

$$\sum_{i=1}^{f} \alpha_i f_i(x) + \sum_{i=1}^{g} \beta_i g_i(x) + \sum_{i=1}^{h} \gamma_i h_i(x) = 0 \tag{1}$$

We substitute the characteristic vector v_s of F_s containing p into Equation (1). Because of the $(x_p - 1)$ factor, we have

$$g_i(v_s) = 0 \quad \text{for all } 1 \le i \le g.$$

Let v_l^c be the characteristic vector of F_l^c. Next, let us consider $h_i(v_s)$:

$$h_i(v_s) = \prod_{j=0}^{|H_i|-1}(w_i \cdot v_s - j) - \sum_{l:p \notin F_l} \frac{\prod_{j=0}^{|H_i|-1}(w_i \cdot v_l^c - j)}{\prod_{j=1}^{k-1}(v_l^c \cdot v_l^c - j)} \prod_{j=1}^{k-1}(v_l^c \cdot v_s - j)$$
$$- \sum_{l:p \in F_l} \frac{\prod_{j=0}^{|H_i|-1}(w_i \cdot v_l - j)}{\prod_{j=1}^{k-1}(v_l \cdot v_l - j)} \prod_{j=1}^{k-1}(v_l \cdot v_s - j).$$

Since $1 \le |F_l \cap F_s| \le k-1$, we have $\prod_{j=1}^{k-1}(v_l \cdot v_s - j) = 0$ except when $s = l$. Since $|F_i| \ge k$ for all i, we have

$$- \sum_{l:p \in F_l} \frac{\prod_{j=0}^{|H_i|-1}(w_i \cdot v_l - j)}{\prod_{j=1}^{k-1}(v_l \cdot v_l - j)} \prod_{j=1}^{k-1}(v_l \cdot v_s - j) = - \frac{\prod_{j=0}^{|H_i|-1}(w_i \cdot v_s - j)}{\prod_{j=1}^{k-1}(v_s \cdot v_s - j)} \prod_{j=1}^{k-1}(v_s \cdot v_s - j)$$
$$= - \prod_{j=0}^{|H_i|-1}(w_i \cdot v_s - j).$$

Since $1 \le |F_l^c \cap F_s| \le k-1$ for $s \ne l$, we have $\prod_{j=1}^{k-1}(v_l^c \cdot v_s - j) = \prod_{j=1}^{k-1}(|F_l^c \cap F_s| - j) = 0$. Thus, we have

$$h_i(v_s) = \prod_{j=0}^{|H_i|-1}(w_i \cdot v_s - j) - \prod_{j=0}^{|H_i|-1}(w_i \cdot v_s - j) = 0 \quad \text{for all } 1 \le i \le h.$$

Finally, we consider $f_i(v_s)$. Since $f_s(v_s) \ne 0$ and $1 \le |F_i \cap F_s| \le k-1$ for $i \ne s$, we get $\alpha_s = 0$ whenever $p \in F_s$.

Next, we substitute the characteristic vector v_s^c of F_s^c into Equation (1), where $p \notin F_s$. Because of the $(x_p - 1)$ factor, we have

$$g_i(v_s^c) = 0 \quad \text{for all } 1 \le i \le g.$$

Next, let us consider $h_i(v_s^c)$. Since $1 \le |F_l^c \cap F_s^c| \le k-1$, we have $\prod_{j=1}^{k-1}(v_l^c \cdot v_s^c - j) = 0$ except when $s = l$. Since $n - |F_i| \ge k$, we have

$$- \sum_{l:p \notin F_l} \frac{\prod_{j=0}^{|H_i|-1}(w_i \cdot v_l^c - j)}{\prod_{j=1}^{k-1}(v_l^c \cdot v_l^c - j)} \prod_{j=1}^{k-1}(v_l^c \cdot v_s^c - j) = - \prod_{j=0}^{|H_i|-1}(w_i \cdot v_s^c - j).$$

Since $1 \leq |F_l \cap F_s^c| \leq k - 1$ for $s \neq l$, we have $\prod_{j=1}^{k-1}(v_l \cdot v_s^c - j) = \prod_{j=1}^{k-1}(|F_l \cap F_s^c| - j) = 0$. Thus, we have

$$h_i(v_s^c) = \prod_{j=0}^{|H_i|-1}(w_i \cdot v_s^c - j) - \prod_{j=0}^{|H_i|-1}(w_i \cdot v_s^c - j) = 0 \quad \text{for all } 1 \leq i \leq h.$$

Finally we consider $f_i(v_s^c)$. Since $1 \leq |F_i \cap F_s^c| \leq k - 1$, by the hypothesis $f_i(v_s^c)$ is also 0 except for $f_s(v_s^c)$. Since $f_s(v_s^c) \neq 0$, we get $\alpha_s = 0$ whenever $p \notin F_s$.

So Equation (1) is reduced to :

$$\sum_{i=1}^{g} \beta_i g_i(x) + \sum_{i=1}^{h} \gamma_i h_i(x) = 0 \tag{2}$$

Next, we substitute the characteristic vector w_s of H_s in order of increasing size into Equation (2). Now we note that $p \in H_s$. Because of the $(x_p - 1)$ factor, we have $g_i(w_s) = 0$ for all $1 \leq i \leq g$. Since the size of H_i is at most $k - 1$ for all i, we have $1 \leq |F_i^c \cap H_s| \leq k - 1$ for $p \in F_i^c$. Thus, the factor $\prod_{j=1}^{k-1}(v_i^c \cdot w_s - j)$ is 0. Similarly, the factor $\prod_{j=1}^{k-1}(v_l \cdot w_s - j)$ is 0 for $p \in F_l$. Thus, we have $h_i(w_s) = \prod_{j=0}^{|H_i|-1}(w_i \cdot w_s - j)$. Since $h_s(w_s) \neq 0$, and $h_i(w_s) = 0$ for $i > s$, we have $\sum_{i=1}^{h} \gamma_i h_i(w_s) = \sum_{i=1}^{s} \gamma_i h_i(w_s)$.

Recall that we substitute the vector w_s in order of increasing size. When we first plug w_1 into Equation (2), we have $\gamma_1 h_1(w_1) = 0$, and thus $\gamma_1 = 0$. Next, we plug w_2 into (2) after dropping $\gamma_1 h_1(w_1)$ term from (2). Then we have $\gamma_2 h_2(w_2) = 0$, and thus $\gamma_2 = 0$. Similarly, we have $\gamma_i = 0$ for all i.

Thus, Equation (1) becomes

$$\sum_{i} \beta_i g_i(x) = 0. \tag{3}$$

Next, we substitute the characteristic vector y_s of G_s in order of increasing size into Equation (3). Thus, we have

$$g_i(y_s) = (y_{s_p} - 1) \prod_{j \in G_i} y_{s_j} = - \prod_{j \in G_i} y_{s_j} \quad \text{for all } 1 \leq i \leq g.$$

Recall that we substitute the vector y_s in order of increasing size. Please note that $g_i(0)$ is the empty product, which is taken to be 1. When we first plug y_1 into Equation (3), we have $g_1(y_1) \neq 0$ and $g_i(y_1) = 0$ for all $i > 1$, and thus $\beta_1 = 0$. Next, we plug y_2 into (3) after dropping $\beta_1 g_1(x)$ term from (3). Then we have $g_2(y_2) \neq 0$ and $g_i(y_2) = 0$ for all $i > 2$, and thus $\beta_2 = 0$. Similarly, we have $\beta_i = 0$ for all i.

This concludes that all the polynomials $f_i(x)$, $g_i(x)$, and $h_i(x)$ are linearly independent. We found $|\mathcal{F}| + |\mathcal{G}| + |\mathcal{H}|$ linearly independent polynomials. All these polynomials are of degree less than or equal to $k - 1$. The space of these multilinear polynomials has dimension $\sum_{i=0}^{k-1} \binom{n}{i}$. We have

$$|\mathcal{F}| + |\mathcal{G}| + |\mathcal{H}| \leq \sum_{i=0}^{k-1} \binom{n}{i}.$$

Since $|\mathcal{G}| = \sum_{i=0}^{k-2} \binom{n-1}{i}$ and $|\mathcal{H}| = \sum_{i=0}^{k-2} \binom{n-1}{i}$, we have $|\mathcal{F}| + 2\sum_{i=0}^{k-2} \binom{n-1}{i} \leq \sum_{i=0}^{k-1} \binom{n}{i}$. This gives us

$$|\mathcal{F}| \leq \binom{n-1}{k-1}$$

finishing the proof of Theorem 6. $\quad \square$

5. Conclusions

We have answered the following question: when is it possible to get the same bound of the Erdős-Ko-Rado theorem for uniform intersecting families in the non-uniform intersecting families?

Since the EKR-type bound for the non-uniform family of subsets of $[n]$, which is $\binom{n-1}{k-1} + \binom{n-1}{k-2} + \cdots + \binom{n-1}{0}$, is much larger than $\binom{n-1}{k-1}$, this question is interesting and deserves further study.

Please note that if we can delete the condition (c) in Theorem 6, we can get the same bound of the Erdős-Ko-Rado theorem for k-uniform intersecting families under the same condition for non-uniform intersecting families of size at least k and at most $n - k$. Another intriguing question motivated by our result is the problem of getting the same bound of Theorem 6 without the condition (c) or finding a better bound for the non-uniform intersecting families than the previous results by the others.

Author Contributions: All authors have contributed equally to this work. All authors have read and agreed to the published version of the manuscript.

Funding: The first author was supported by Basic Science Research Program through the National Research Foundation of Korea (NRF) funded by the Ministry of Education, Science and Technology (2011-0025252). The second author was supported by Basic Science Research Program through the National Research Foundation of Korea(NRF) funded by the Ministry of Education (2017R1A6A3A04005963).

Acknowledgments: All authors sincerely appreciate the reviewers for their valuable comments and suggestions to improve the paper.

Conflicts of Interest: The authors declare no conflict of interest.

References

1. Erdős, P.; Ko, C.; Rado, R. Intersection theorem for systems of finite sets. *Q. J. Math. Oxf. Ser.* **1961**, *12*, 313–320. [CrossRef]
2. Frankl, P. The Erdős-Ko-Rado theorem is true for $n = ckt$. In *Combinatorics, Proceedings of the Fifth Hungarian Colloquium, Keszthely*; North-Holland Publishing Company: Amsterdam, The Netherlands, 1978; Volume 1, pp. 365–375.
3. Wilson, R. The exact bound in the Erdős-Ko-Rado theorem. *Combinatorica* **1984**, *4*, 247–257. [CrossRef]
4. Alon, N.; Aydinian, H.; Huang, H. Maximizing the number of nonnegative subsets. *SIAM J. Discret. Math.* **2014**, *28*, 811–816. [CrossRef]
5. Deza, M.; Frankl, P. Erdős-Ko-Rado theorem–22 years later. *SIAM J. Algebr. Discret. Methods* **1983**, *4*, 419–431. [CrossRef]
6. Ray-Chaudhuri, D.; Wilson, R. On t-designs. *Osaka J. Math.* **1975**, *12*, 737–744.
7. Frankl, P.; Wilson, R. Intersection theorems with geometric consequences. *Combinatorica* **1981**, *1*, 357–368. [CrossRef]
8. Deza, M.; Frankl, P.; Singhi, N. On functions of strength t. *Combinatorica* **1983**, *3*, 331–339. [CrossRef]
9. Alon, N.; Babai, L.; Suzuki, H. Multilinear polynomials and Frankl-Ray-Chaudhuri-Wilson type intersection theorems. *J. Comb. Theory Ser. A* **1991**, *58*, 165–180. [CrossRef]
10. Hwang, K.; Kim, Y. A proof of Alon-Babai-Suzuki's Conjecture and Multilinear Polynomials. *Eur. J. Comb.* **2015**, *43*, 289–294. [CrossRef]
11. Blokhuis, A. Solution of an extremal problem for sets using resultants of polynomials. *Combinatorica* **1990**, *10*, 393–396. [CrossRef]
12. Chen, W.Y.C.; Liu, J. Set systems with L-intersections modulo a prime number. *J. Comb. Theory Ser. A* **2009**, *116*, 120–131. [CrossRef]
13. Füredi, Z.; Hwang, K.; Weichsel, P. A proof and generalization of the Erős-Ko-Rado theorem using the method of linearly independent polynomials. In *Topics in Discrete Mathematics*; Algorithms Combin. 26; Springer: Berlin, Germnay, 2006; pp. 215–224.
14. Liu, J.; Yang, W. Set systems with restricted k-wise L-intersections modulo a prime number. *Eur. J. Comb.* **2014**, *36*, 707–719. [CrossRef]
15. Qian, J.; Ray-Chaudhuri, D. On mod p Alon-Babai-Suzuki inequality. *J. Algebr. Comb.* **2000**, *12*, 85–93. [CrossRef]
16. Ramanan, G. Proof of a conjecture of Frankl and Füredi. *J. Comb. Theory Ser. A* **1997**, *79*, 53–67. [CrossRef]
17. Snevily, H. On generalizations of the de Bruijn-Erdős theorem. *J. Comb. Theory Ser. A* **1994**, *68*, 232–238. [CrossRef]

18. Snevily, H. A sharp bound for the number of sets that pairwise intersect at k positive values. *Combinatorica* **2003**, *23*, 527–532. [CrossRef]
19. Wang, X.; Wei, H.; Ge, G. A strengthened inequality of Alon-Babai-Suzuki's conjecture on set systems with restricted intersections modulo p. *Discret. Math.* **2018**, *341*, 109–118. [CrossRef]

Article

A Note on Parametric Kinds of the Degenerate Poly-Bernoulli and Poly-Genocchi Polynomials

Taekyun Kim [1], Waseem A. Khan [2], Sunil Kumar Sharma [3,*] and Mohd Ghayasuddin [4]

[1] Department of Mathematics, Kwangwoon University, Seoul 139-701, Korea; tkkim@kw.ac.kr
[2] Department of Mathematics and Natural Sciences, Prince Mohammad Bin Fahd University, P.O Box 1664, Al Khobar 31952, Kingdom of Saudi Arabia; wkhan1@pmu.edu.sa
[3] College of Computer and Information Sciences, Majmaah University, Majmaah 11952, Saudi Arabia
[4] Department of Mathematics, Integral University Campus, Shahjahanpur 242001, India; ghayas.maths@gmail.com
* Correspondence: s.sharma@mu.edu.sa

Received: 20 March 2020; Accepted: 2 April 2020; Published: 13 April 2020

Abstract: Recently, the parametric kind of some well known polynomials have been presented by many authors. In a sequel of such type of works, in this paper, we introduce the two parametric kinds of degenerate poly-Bernoulli and poly-Genocchi polynomials. Some analytical properties of these parametric polynomials are also derived in a systematic manner. We will be able to find some identities of symmetry for those polynomials and numbers.

Keywords: degenerate poly-Bernoulli polynomials; degenerate poly-Genocchi polynomials; stirling numbers

1. Introduction

Special functions, polynomials and numbers play a prominent role in the study of many areas of mathematics, physics and engineering. In particular, the Appell polynomials and numbers are frequently used in the development of pure and applied mathematics related to functional equations in differential equations, approximation theories, interpolation problems, summation methods, quadrature rules and their multidimensional extensions (see [1]).The sequence of Appell polynomials $A_j(z)$ can be signified as follows:

$$\frac{d}{dz}A_j(z) = jA_{j-1}(z), \quad A_0(z) \neq 0, z = \eta + i\xi \in \mathbb{C}, \quad j \in \mathbb{N}, \tag{1}$$

or equivalently

$$A(z)e^{\eta z} = \sum_{j=0}^{\infty} A_j(\eta)\frac{z^j}{j!}, \tag{2}$$

where

$$A(z) = A_0 + A_1\frac{z}{1!} + A_2\frac{z^2}{2!} + \cdots + A_j\frac{z^j}{j!} + \cdots, \quad A_0 \neq 0,$$

is a formal power series with coefficients A_j known as Appell numbers.

The well known degenerate exponential function is defined by (see [2])

$$e_\mu^\eta(z) = (1 + \mu z)^{\frac{\eta}{\mu}}, \quad e_\mu(z) = e_\mu^1(z), (\mu \in \mathbb{R}). \tag{3}$$

In 1956 and 1979, Carlitz [3,4] introduced and investigated the following degenerate Bernoulli and Euler polynomials:

$$\frac{z}{e_\mu(z)-1}e_\mu^\eta(z) = \frac{z}{(1+\mu z)^{\frac{1}{\mu}}-1}(1+\mu z)^{\frac{\eta}{\mu}} = \sum_{s=0}^{\infty}\beta_s(\eta;\mu)\frac{z^s}{s!}, \tag{4}$$

and

$$\frac{2}{e_\mu(z)+1}e_\mu^\eta(z) = \frac{2}{(1+\mu z)^{\frac{1}{\mu}}-1}(1+\mu z)^{\frac{\eta}{\mu}} = \sum_{s=0}^{\infty}\mathfrak{E}_s(\eta;\mu)\frac{z^s}{s!}. \tag{5}$$

Note that

$$\lim_{\mu\longrightarrow 0}\beta_s(\eta;\mu) = B_s(\eta), \quad \lim_{\mu\longrightarrow 0}\mathfrak{E}_s(\eta;\mu) = E_s(\eta),$$

where $B_s(\eta)$ and $E_s(\eta)$ are the classical Bernoulli and Euler polynomials (see [5,6]).

Lim [7] introduced the degenerate Genocchi polynomials $G_j^{(p)}(\eta;\mu)$ of order p by means of the undermentioned generating function:

$$\left(\frac{2z}{e_\mu(z)+1}\right)^p e_\mu^\eta(z) = \left(\frac{2z}{(1+\mu z)^{\frac{1}{\mu}}-1}\right)^p (1+\mu z)^{\frac{\eta}{\mu}} = \sum_{j=0}^{\infty}G_j^{(p)}(\eta;\mu)\frac{z^j}{j!}, \tag{6}$$

so that

$$G_j^{(p)}(\eta;\mu) = \sum_{s=0}^{j}\binom{j}{s}G_s^{(p)}(\mu)\left(\frac{\eta}{\mu}\right)_{j-s}. \tag{7}$$

From Equation (6), we note that

$$\lim_{\mu\longrightarrow 0}\sum_{s=0}^{\infty}G_j^{(p)}(\eta;\mu)\frac{z^j}{j!} = \lim_{\mu\longrightarrow 0}\left(\frac{2z}{(1+\mu z)^{\frac{1}{\mu}}-1}\right)^p(1+\mu z)^{\frac{\eta}{\mu}}$$

$$= \left(\frac{2z}{e^z+1}\right)^p e^{\eta z} = \sum_{j=0}^{\infty}G_j^{(p)}(\eta)\frac{z^j}{j!},$$

where $G_j^{(p)}(\eta)$ are the generalized Genocchi polynomials of order p (see [8–11]).

The degenerate poly-Bernoulli and poly-Genocchi polynomials are defined by (see [12–14])

$$\frac{\mathrm{Li}_k(1-e^{-z})}{e_\mu(z)-1}e_\mu^\eta(z) = \frac{\mathrm{Li}_k(1-e^{-z})}{(1+\mu z)^{\frac{1}{\mu}}-1}(1+\mu z)^{\frac{\eta}{\mu}} = \sum_{s=0}^{\infty}B_s^{(k)}(\eta;\mu)\frac{z^s}{s!}, (k\in\mathbb{Z}), \tag{8}$$

and

$$\frac{2\mathrm{Li}_k(1-e^{-z})}{e_\mu(z)+1}e_\mu^\eta(z) = \frac{2\mathrm{Li}_k(1-e^{-z})}{(1+\mu z)^{\frac{1}{\lambda}}+1}(1+\mu z)^{\frac{\eta}{\mu}} = \sum_{s=0}^{\infty}G_s^{(k)}(\eta;\mu)\frac{z^s}{s!}, (k\in\mathbb{Z}). \tag{9}$$

Here, we note that (see [5,15]).

$$\lim_{\mu\longrightarrow 0}B_s^{(k)}(\eta;\mu) = B_s^{(k)}(\eta), \quad \lim_{\mu\longrightarrow 0}G_s^{(k)}(\eta;\mu) = G_s^{(k)}(\eta),$$

The Stirling numbers of the first kind are given by (see, [16–18])

$$(a)_s = a(a-1)\cdots(a-s+1) = \sum_{k=0}^{s} S^{(1)}(s,k)a^k, (k \geq 0), \tag{10}$$

and the Stirling numbers of the second kind are defined by (see [19,20])

$$a^s = \sum_{k=0}^{s} S^{(2)}(k,s)(a)_k. \tag{11}$$

The degenerate Stirling numbers of the of the second kind are defined by (see [10,21,22])

$$\frac{1}{k!}(e_\mu(t)-1)^k = \sum_{k=s}^{\infty} S^{(2)}_\mu(k,s)\frac{t^k}{k!}, (k \geq 0). \tag{12}$$

Note that $\lim_{\mu\longrightarrow 0} S^{(2)}_\mu(k,s) = S^{(2)}(k,s), (s,k \geq 0)$.

In the year (2017, 2018), Jamei et al. [23,24] introduced the two parametric kinds of exponential functions as follows (see also [6,23–25]):

$$e^{\eta z}\cos\xi z = \sum_{k=0}^{\infty} C_k(\eta,\xi)\frac{z^k}{k!}, \tag{13}$$

and

$$e^{\eta z}\sin\xi z = \sum_{k=0}^{\infty} S_k(\eta,\xi)\frac{z^k}{k!}, \tag{14}$$

where

$$C_k(\eta,\xi) = \sum_{j=0}^{[\frac{k}{2}]} \binom{k}{2j} (-1)^j \eta^{k-2j}\xi^{2j}, \tag{15}$$

and

$$S_k(\eta,\xi) = \sum_{j=0}^{[\frac{k-1}{2}]} \binom{k}{2j+1} (-1)^j \eta^{k-2j-1}\xi^{2j+1}. \tag{16}$$

Recently, Kim et al. [2] introduced the following degenerate type parametric exponential functions:

$$e_\mu^\eta(z)\cos_\mu^\xi(z) = \sum_{k=0}^{\infty} C_{k,\mu}(\eta,\xi)\frac{z^k}{k!}, \tag{17}$$

and

$$e_\mu^\eta(z)\sin_\mu^\xi(z) = \sum_{k=0}^{\infty} S_{k,\mu}(\eta,\xi)\frac{z^k}{k!}, \tag{18}$$

where

$$C_{r,\mu}(\eta,\xi) = \sum_{k=0}^{[\frac{r}{2}]} \sum_{q=2k}^{r} \binom{r}{q} (-1)^k \mu^{q-2k}\xi^{2k} S^1(q,2k)(\eta)_{r-q,\mu}, \tag{19}$$

and

$$S_{r,\mu}(\eta,\xi) = \sum_{k=0}^{[\frac{r-1}{2}]} \sum_{q=2k+1}^{r} \binom{r}{q} (-1)^k \mu^{q-2k-1}\xi^{2k+1} S^1(q,2k+1)(\eta)_{r-q,\mu}. \tag{20}$$

Motivated by the importance and potential applications in certain problems in number theory, combinatorics, classical and numerical analysis and physics, several families of degenerate Bernoulli and Euler polynomials and degenerate versions of special polynomials have been recently studied by many authors, (see [3–5,11–13,16]). Recently, Kim and Kim [2] have introduced the degenerate Bernoulli and degenerate Euler polynomials of a complex variable. By separating the real and imaginary parts, they introduced the parametric kinds of these degenerate polynomials.

The main object of this article is to present the parametric kinds of degenerate poly-Bernoulli and poly-Genocchi polynomials in terms of the degenerate type parametric exponential functions. We also investigate some fundamental properties of our introduced parametric polynomials.

2. Parametric Kinds of the Degenerate Poly-Bernoulli Polynomials

In this section, we define the two parametric kinds of degenerate poly-Bernoulli polynomials by means of the two special generating functions involving the degenerate exponential as well as trigonometric functions.

It is well known that (see [2])

$$e^{(\eta+i\xi)z} = e^{\eta z}e^{i\xi z} = e^{\eta z}(\cos \xi z + i \sin \xi z), \tag{21}$$

The degenerate trigonometric functions are defined by (see [19])

$$\cos_\mu z = \frac{e_\mu^i(z) + e_\mu^{-i}(z)}{2}, \quad \sin_\mu z = \frac{e_\mu^i(z) - e_\mu^{-i}(z)}{2i}. \tag{22}$$

Note that, we have

$$\lim_{\mu \to 0} \cos_\mu z = \cos z, \quad \lim_{\mu \to 0} \sin_\mu z = \sin z.$$

In view of Equation (8), we have

$$\frac{\mathrm{Li}_k(1 - e^{-z})}{e_\mu(z) - 1}e_\mu^{\eta+i\xi}(z) = \sum_{j=0}^{\infty} B_{j,\mu}^{(k)}(\eta + i\xi)\frac{z^j}{j!}, \tag{23}$$

and

$$\frac{\mathrm{Li}_k(1 - e^{-z})}{e_\mu(z) - 1}e_\mu^{\eta-i\xi}(z) = \sum_{j=0}^{\infty} B_{j,\mu}^{(k)}(\eta - i\xi)\frac{z^j}{j!}. \tag{24}$$

From Equations (23) and (24), we note that

$$\frac{\mathrm{Li}_k(1 - e^{-z})}{e_\mu(z) - 1}e_\mu^{\eta}(z)\cos_\mu^{\xi}(z) = \sum_{j=0}^{\infty} \left(\frac{B_{j,\mu}^{(k)}(\eta + i\xi) + B_{j,\mu}^{(k)}(\eta - i\xi)}{2}\right)\frac{z^j}{j!}, \tag{25}$$

and

$$\frac{\mathrm{Li}_k(1 - e^{-z})}{e_\mu(z) - 1}e_\mu^{\eta}(z)\sin_\mu^{\xi}(z) = \sum_{j=0}^{\infty} \left(\frac{B_{j,\mu}^{(k)}(\eta + i\xi) - B_{j,\mu}^{(k)}(\eta - i\xi)}{2i}\right)\frac{z^j}{j!}. \tag{26}$$

Definition 1. *The degenerate cosine-poly-Bernoulli polynomials* $B_{p,\mu}^{(k,c)}(\eta, \xi)$ *and degenerate sine-poly-Bernoulli polynomials* $B_{p,\mu}^{(k,s)}(\eta, \xi)$ *for nonnegative integer p are defined, respectively, by*

$$\frac{\mathrm{Li}_k(1 - e^{-z})}{e_\mu(z) - 1}e_\mu^{\eta}(z)\cos_\mu^{\xi}(z) = \sum_{p=0}^{\infty} B_{p,\mu}^{(k,c)}(\eta, \xi)\frac{z^p}{p!}, \tag{27}$$

and

$$\frac{\mathrm{Li}_k(1 - e^{-z})}{e_\mu(z) - 1}e_\mu^{\eta}(z)\sin_\mu^{\xi}(z) = \sum_{p=0}^{\infty} B_{p,\mu}^{(k,s)}(\eta, \xi)\frac{z^p}{p!}. \tag{28}$$

For $\eta = \xi = 0$ *in Equations (27) and (28), we get*

$$B_{p,\mu}^{(k,c)}(0,0) = B_{p,\mu}^{(k)}, B_{p,\mu}^{(k,s)}(0,0) = 0, (p \geq 0).$$

Note that $\lim_{\mu \to 0} B_{p,\mu}^{(k,c)}(\eta,\xi) = B_p^{(k,c)}(\eta,\xi)$, $\lim_{\mu \to 0} B_{p,\mu}^{(k,s)}(\eta,\xi) = B_p^{(k,s)}(\eta,\xi)$, $(p \geq 0)$, where $B_p^{(k,c)}(\eta,\xi)$ and $B_p^{(k,s)}(\eta,\xi)$ are the new type of poly-Bernoulli polynomials.

Based on Equations (25)–(28), we determine

$$B_{p,\mu}^{(k,c)}(\eta,\xi) = \frac{B_{p,\mu}^{(k)}(\eta + i\xi) + B_{p,\mu}^{(k)}(\eta - i\xi)}{2}, \tag{29}$$

and

$$B_{p,\mu}^{(k,s)}(\eta,\xi) = \frac{B_{p,\mu}^{(k)}(\eta + i\xi) - B_{p,\mu}^{(k)}(\eta - i\xi)}{2i}. \tag{30}$$

Theorem 1. *Let $k \in \mathbb{Z}$ and $j \geq 0$. Then*

$$B_{j,\mu}^{(k)}(\eta + i\xi) = \sum_{q=0}^{j} \binom{j}{q} B_{j-q,\mu}^{(k)}(\eta)(i\xi)_{q,\mu}$$

$$= \sum_{q=0}^{j} \binom{j}{q} B_{j-q,\mu}^{(k)}(\eta + i\xi)_{q,\mu}, \tag{31}$$

and

$$B_{j,\mu}^{(k)}(\eta - i\xi) = \sum_{q=0}^{j} \binom{j}{q} B_{j-q,\mu}^{(k)}(\eta)(-1)^q(i\xi)_{q,\mu}$$

$$= \sum_{q=0}^{j} \binom{j}{q} B_{j-q,\mu}^{(k)}(\eta - i\xi)_{q,\mu}. \tag{32}$$

Proof. From Equation (23), we have

$$\sum_{j=0}^{\infty} B_{j,\mu}^{(k)}(\eta + i\xi)\frac{z^j}{j!} = \frac{\mathrm{Li}_k(1 - e^{-z})}{e_\mu(z) - 1}e_\mu^\eta(z)e_\mu^{i\xi}(z)$$

$$= \left(\sum_{j=0}^{\infty} B_{j,\mu}^{(k)}(\eta)\frac{z^j}{j!}\right)\left(\sum_{q=0}^{\infty}(i\xi)_{q,\mu}\frac{z^q}{q!}\right)$$

$$= \sum_{j=0}^{\infty}\left(\sum_{q=0}^{j}\binom{j}{q}B_{j-q,\mu}^{(k)}(\eta)(i\xi)_{q,\mu}\right)\frac{z^j}{j!}. \tag{33}$$

Similarly, we find

$$\frac{\mathrm{Li}_k(1 - e^{-z})}{e_\mu(z) - 1}e_\mu^\eta(z)e_\mu^{i\xi}(z) = \left(\sum_{j=0}^{\infty} B_{j,\mu}^{(k)}\frac{z^j}{j!}\right)\left(\sum_{q=0}^{\infty}(\eta + i\xi)_{q,\mu}\frac{z^q}{q!}\right)$$

$$= \sum_{j=0}^{\infty}\left(\sum_{q=0}^{j}\binom{j}{q}B_{j-q,\mu}^{(k)}(\eta + i\xi)_{q,\mu}\right)\frac{z^j}{j!}. \tag{34}$$

In view of Equations (33) and (34), we obtain our first claimed result shown in Equation (31). Similarly, we can establish our second result shown in Equation (32). $\square$

Theorem 2. *The following results hold true:*

$$
B_{j,\mu}^{(k,c)}(\eta,\xi) = \sum_{r=0}^{j} \binom{j}{r} B_{r,\mu}^{(k)} C_{j-r,\mu}(\eta,\xi)
$$

$$
= \sum_{r=0}^{\left[\frac{q}{2}\right]} \sum_{q=2r}^{j} \binom{j}{q} \mu^{q-2r}(-1)^r \xi^{2r} S^{(1)}(q,2r) B_{j-q,\mu}^{(k)}(\eta),
\tag{35}
$$

and

$$
B_{j,\mu}^{(k,s)}(\eta,\xi) = \sum_{r=0}^{j} \binom{j}{r} B_{r,\mu}^{(k)} S_{j-r,\mu}(\eta,\xi)
$$

$$
= \sum_{r=0}^{\left[\frac{q-1}{2}\right]} \sum_{q=2r+1}^{j} \binom{j}{q} \mu^{q-2r-1}(-1)^r \xi^{2r+1} S^{(1)}(q,2r+1) B_{j-q,\mu}^{(k)}(\eta).
\tag{36}
$$

Proof. From Equations (27) and (17), we see

$$
\sum_{j=0}^{\infty} B_{j,\mu}^{(k,c)}(\eta,\xi) \frac{z^j}{j!} = \frac{\mathrm{Li}_k(1-e^{-z})}{e_\mu(z)-1} e_\mu^\eta(z) \cos_\mu^\xi(z)
$$

$$
= \left(\sum_{r=0}^{\infty} B_{r,\mu}^{(k)} \frac{z^r}{r!} \right) \left(\sum_{j=0}^{\infty} C_{j,\mu}(\eta,\xi) \frac{z^j}{j!} \right)
$$

$$
= \sum_{j=0}^{\infty} \left(\sum_{r=0}^{j} \binom{j}{r} B_{r,\mu}^{(k)} C_{j-r,\mu}(\eta,\xi) \right) \frac{z^j}{j!}.
\tag{37}
$$

Now, by using Equations (27) and (10), we find

$$
\frac{\mathrm{Li}_k(1-e^{-z})}{e_\mu(z)-1} e_\mu^\eta(z) \cos_\mu^\xi(z) = \sum_{j=0}^{\infty} B_{j,\mu}^{(k)}(\eta) \frac{z^j}{j!} \sum_{p=0}^{\infty} \sum_{r=0}^{\left[\frac{q}{2}\right]} \mu^{l-2r}(-1)^r y^{2r} S^{(1)}(q,2r) \frac{z^r}{r!}
$$

$$
= \sum_{j=0}^{\infty} \left(\sum_{q=0}^{j} \sum_{r=0}^{\left[\frac{q}{2}\right]} \binom{j}{q} \mu^{q-2r}(-1)^r \xi^{2r} S^{(1)}(q,2r) B_{j-r,\mu}^{(k)}(\eta) \right) \frac{z^j}{j!}
$$

$$
= \sum_{j=0}^{\infty} \left(\sum_{r=0}^{\left[\frac{q}{2}\right]} \sum_{q=2r}^{j} \binom{j}{q} \mu^{q-2r}(-1)^r \xi^{2r} S^{(1)}(q,2r) B_{j-q,\mu}^{(k)}(\eta) \right) \frac{z^j}{j!}.
\tag{38}
$$

Therefore, from Equations (37) and (38), we attain our needed result, Equation (35). Similarly, we can obtain Equation (36). $\square$

Theorem 3. *Each of the following identities holds true:*

$$
B_{r,\mu}^{(2,c)}(\eta,\xi) = \sum_{q=0}^{r} \binom{r}{q} \frac{q! B_q}{q+1} B_{r-q,\mu}^{(c)}(\eta,\xi),
\tag{39}
$$

and

$$B_{r,\mu}^{(2,s)}(\eta,\xi) = \sum_{q=0}^{r} \binom{r}{q} \frac{q! B_q}{q+1} B_{r-q,\mu}^{(s)}(\eta,\xi). \tag{40}$$

Proof. In view of Equation (27), we have

$$\sum_{r=0}^{\infty} B_{r,\mu}^{(k,c)}(\eta,\xi) \frac{z^r}{r!} = \frac{\mathrm{Li}_k(1-e^{-z})}{e_\mu(z)-1} e_\mu^\eta(z) \cos_\mu^\xi(z)$$

$$= \frac{e_\mu^\eta(z) \cos_\mu^\xi(z)}{e_\mu(z)-1} \int_0^z \underbrace{\frac{1}{e^u-1} \int_0^u \frac{1}{e^u-1} \cdots \frac{1}{e^u-1} \int_0^u \frac{u}{e^u-1} du \cdots du}_{(k-1)-\text{times}}. \tag{41}$$

Upon setting $k = 2$, we obtain

$$\sum_{r=0}^{\infty} B_{r,\mu}^{(2,c)}(\eta,\xi) \frac{z^r}{r!} = \frac{e_\mu^\eta(z) \cos_\mu^\xi(z)}{e_\mu(z)-1} \int_0^z \frac{u}{e^u-1} du$$

$$= \left(\sum_{q=0}^{\infty} \frac{B_q z^q}{(q+1)} \right) \frac{e_\mu^\eta(z) \cos_\mu^\xi(z)}{e_\mu(z)-1}$$

$$= \left(\sum_{q=0}^{\infty} \frac{q! B_q z^q}{(q+1) q!} \right) \left(\sum_{r=0}^{\infty} B_{r,\mu}^{(c)}(\eta,\xi) \frac{z^r}{r!} \right).$$

$$= \sum_{r=0}^{\infty} \sum_{q=0}^{r} \binom{r}{q} \frac{q! B_q}{q+1} B_{r-q,\mu}^{(c)}(\eta,\xi) \frac{z^r}{r!},$$

which gives our required result, Equation (39). The proof of Equation (40) is similar; therefore, we omit the proof. $\square$

Theorem 4. *Let* $k \in \mathbb{Z}$, *then*

$$B_{j,\mu}^{(k,c)}(\eta,\xi) = \sum_{r=0}^{j} \binom{j}{r} \left(\sum_{q=1}^{r+1} \frac{(-1)^{q+r+1} l! S_2(r+1,q)}{q^k(r+1)} \right) B_{j-r,\mu}^{(c)}(\eta,\xi), \tag{42}$$

and

$$B_{j,\mu}^{(k,s)}(\eta,\xi) = \sum_{r=0}^{j} \binom{j}{r} \left(\sum_{q=1}^{r+1} \frac{(-1)^{q+r+1} q! S_2(r+1,q)}{q^k(r+1)} \right) B_{j-r,\mu}^{(s)}(\eta,\xi). \tag{43}$$

Proof. From Equations (27) and (11), we see

$$\sum_{j=0}^{\infty} B_{j,\mu}^{(k,c)}(\eta,\xi) \frac{z^j}{j!} = \left(\frac{\mathrm{Li}_k(1-e^{-z})}{z} \right) \left(\frac{z e_\mu^\eta(z) \cos_\mu^\xi(z)}{e_\mu(z)-1} \right). \tag{44}$$

Now

$$\frac{1}{z} \mathrm{Li}_k(1-e^{-z}) = \frac{1}{z} \sum_{q=1}^{\infty} \frac{(1-e^{-z})^q}{q^k}$$

$$= \frac{1}{z} \sum_{q=1}^{\infty} \frac{(-1)^q}{q^k} q! \sum_{r=l}^{\infty} (-1)^r S_2(r,q) \frac{z^r}{r!}$$

$$= \frac{1}{z} \sum_{r=q}^{\infty} \sum_{q=1}^{r} \frac{(-1)^{q+r}}{q^k} q! S_2(r,q) \frac{z^r}{r!}$$

$$= \sum_{r=0}^{\infty} \left(\sum_{q=1}^{q+1} \frac{(-1)^{q+r+1}}{q^k} l! \frac{S_2(r+1,q)}{r+1} \right) \frac{z^r}{r!}. \tag{45}$$

On using Equation (45) in (44), we find

$$\sum_{j=0}^{\infty} B_{j,\mu}^{(k,c)}(\eta,\xi)\frac{z^j}{j!} = \sum_{r=0}^{\infty} \left(\sum_{q=1}^{r+1} \frac{(-1)^{q+r+1}}{q^k} l! \frac{S_2(r+1,q)}{r+1} \right) \frac{z^r}{r!} \left(\sum_{j=0}^{\infty} B_{j,\mu}^{(c)}(\eta,\xi)\frac{z^j}{j!} \right).$$

Replacing j by $j-r$ in the right side of above expression and after equating the coefficients of z^j, we obtain our needed result, Equation (42). Similarly, we can derive our second result, Equation (43). $\square$

Theorem 5. *The following recurrence relation holds true:*

$$B_{j,\mu}^{(k,c)}(\eta+1,\xi) - B_{j,\mu}^{(k,c)}(\eta,\xi)$$

$$= \sum_{r=1}^{j} \binom{j}{r} \left(\sum_{q=0}^{r-1} \frac{(-1)^{q+r+1}}{(q+1)^k}(q+1)!S_2(r,q+1) \right) C_{j-r,\mu}(\eta,\xi), \tag{46}$$

and

$$B_{j,\mu}^{(k,s)}(\eta+1,\xi) - B_{j,\mu}^{(k,s)}(\eta,\xi)$$

$$= \sum_{r=1}^{j} \binom{j}{r} \left(\sum_{q=0}^{r-1} \frac{(-1)^{q+r+1}}{(q+1)^k}(q+1)!S_2(r,q+1) \right) S_{j-r,\mu}(\eta,\xi). \tag{47}$$

Proof. In view of Equation (27), we have

$$\sum_{j=0}^{\infty} B_{j,\mu}^{(k,c)}(\eta+1,\xi)\frac{z^j}{j!} - \sum_{j=0}^{\infty} B_{j,\mu}^{(k,c)}(\eta,\xi)\frac{z^j}{j!}$$

$$= \frac{\mathrm{Li}_k(1-e^{-z})}{e_\mu(z)-1}e_\mu^{(\eta+1)}(z)\cos_\mu^{\xi}(z) - \frac{\mathrm{Li}_k(1-e^{-z})}{e_\mu(z)-1}e_\mu^{(\eta)}(z)\cos_\mu^{\xi}(z)$$

$$= \mathrm{Li}_k(1-e^{-z})e_\mu^{(\eta)}(z)\cos_\mu^{\xi}(z)$$

$$= \sum_{q=0}^{\infty} \frac{(1-e^{-z})^{q+1}}{(q+1)^k}e_\mu^{(\eta)}(z)\cos_\mu^{\xi}(z)$$

$$= \sum_{r=1}^{\infty} \left(\sum_{q=0}^{r-1} \frac{(-1)^{q+r+1}}{(q+1)^k}(q+1)!S_2(r,q+1) \right) \frac{z^r}{r!}e_\mu^{(\eta)}(z)\cos_\mu^{\xi}(z)$$

$$= \left(\sum_{r=1}^{\infty} \left(\sum_{q=0}^{r-1} \frac{(-1)^{q+r+1}}{(q+1)^k}(q+1)!S_2(r,q+1) \right) \frac{z^r}{r!} \right) \left(\sum_{j=0}^{\infty} C_{j,\mu}(\eta,\xi)\frac{z^j}{j!} \right),$$

which upon replacing j by $j-r$ in the right side of above expression and after equating the coefficients of z^j, yields our first claimed result, Equation (46). Similarly, we can establish our second result, Equation (47). $\square$

Theorem 6. *Let $k \in \mathbb{Z}$ and $j \geq 0$, then we have*

$$B_{j,\mu}^{(k,c)}(\eta+\gamma,\xi) = \sum_{r=0}^{j} \binom{j}{r} B_{j-r,\mu}^{(k,c)}(\eta,\xi)(\gamma)_{r,\mu}, \tag{48}$$

and

$$B_{j,\mu}^{(k,s)}(\eta+\gamma,\xi) = \sum_{r=0}^{j} \binom{j}{r} B_{j-r,\mu}^{(k,s)}(\eta,\xi)(\gamma)_{r,\mu}. \tag{49}$$

Proof. On using Equation (27), we find

$$\sum_{j=0}^{\infty} B_{j,\mu}^{(k,c)}(\eta+\gamma,\xi)\frac{z^j}{j!} = \frac{\mathrm{Li}_k(1-e^{-z})}{e_\mu(z)-1}e_\mu^{(\eta+\gamma)}(z)\cos_\mu^{\xi}(z)$$

$$= \left(\sum_{j=0}^{\infty} B_{j,\mu}^{(k,c)}(\eta,\xi)\frac{z^j}{j!}\right)\left(\sum_{r=0}^{\infty}(\gamma)_{r,\mu}\frac{z^r}{r!}\right)$$

$$= \sum_{j=0}^{\infty}\left(\sum_{r=0}^{j}\binom{j}{r}B_{j-r,\mu}^{(k,c)}(\eta,\xi)(\gamma)_{r,\mu}\right)\frac{z^j}{j!}.$$

By comparing the coefficients of z^j on both sides, we obtain the result, Equation (48). The proof of Equation (49) is similar to Equation (48).

□

Theorem 7. *If $k \in \mathbb{Z}$ and $j \geq 0$, then*

$$B_{j,\mu}^{(k,c)}(\eta,\xi) = \sum_{r=0}^{j}\sum_{q=0}^{r}\binom{j}{r}(\eta)_q\, S_\mu^{(2)}(r,q)B_{j-r,\mu}^{(k,c)}(0,\xi), \tag{50}$$

and

$$B_{j,\mu}^{(k,s)}(\eta,\xi) = \sum_{r=0}^{j}\sum_{q=0}^{r}\binom{j}{r}(\eta)_q\, S_\mu^{(2)}(r,q)B_{j-r,\mu}^{(k,s)}(0,\xi). \tag{51}$$

Proof. From Equations (27) and (12), we find

$$\sum_{j=0}^{\infty} B_{j,\mu}^{(k,c)}(\eta,\xi)\frac{z^j}{j!} = \frac{\mathrm{Li}_k(1-e^{-z})}{e_\mu(z)-1}(e_\mu(z)-1+1)^\eta\cos_\mu^{\xi}(z)$$

$$= \frac{\mathrm{Li}_k(1-e^{-z})}{e_\mu(z)-1}\sum_{q=0}^{\infty}\binom{\eta}{q}(e_\mu(z)-1)^q\cos_\mu^{\xi}(z)$$

$$= \frac{\mathrm{Li}_k(1-e^{-z})}{e_\mu(z)-1}\cos_\mu^{\xi}(z)\sum_{q=0}^{\infty}(\eta)_q\sum_{r=q}^{\infty}S_\mu^{(2)}(r,q)\frac{z^r}{r!}$$

$$= \sum_{j=0}^{\infty} B_{j,\mu}^{(k,c)}(0,\xi)\frac{z^j}{j!}\sum_{r=0}^{\infty}\left(\sum_{q=0}^{r}(\eta)_q S_\mu^{(2)}(r,q)\right)\frac{z^r}{r!}$$

$$= \sum_{j=0}^{\infty}\left(\sum_{r=0}^{j}\sum_{q=0}^{r}\binom{j}{r}(\eta)_q S_\mu^{(2)}(r,q B_{j-r,\mu}^{(k,c)}(0,\xi)\right)\frac{z^j}{j!}.$$

On comparing the coefficients of z^j on both sides, we obtain our required result, Equation (50). The proof of Equation (51) is similar to Equation (50).

□

3. Parametric Kinds of Degenerate Poly-Genocchi Polynomials

In this section, we introduce the two parametric kinds of degenerate poly-Genocchi polynomials by defining the two special generating functions involving the degenerate exponential as well as trigonometric functions.

In view of Equation (9), we have

$$\frac{2\mathrm{Li}_k(1-e^{-z})}{e_\mu(z)+1}e_\mu^{\eta+i\xi}(z) = \sum_{j=0}^\infty G_{j,\mu}^{(k)}(\eta+i\xi)\frac{z^j}{j!}, \tag{52}$$

and

$$\frac{2\mathrm{Li}_k(1-e^{-z})}{e_\mu(z)+1}e_\mu^{\eta-i\xi}(z) = \sum_{j=0}^\infty G_{j,\mu}^{(k)}(\eta-i\xi)\frac{z^j}{j!}. \tag{53}$$

From Equations (52) and (53), we can easily get

$$\frac{2\mathrm{Li}_k(1-e^{-z})}{e_\mu(z)+1}e_\mu^{\eta}(z)\cos_\mu^\xi(z) = \sum_{j=0}^\infty \left(\frac{G_{j,\mu}^{(k)}(\eta+i\xi) + G_{j,\mu}^{(k)}(\eta-i\xi)}{2} \right)\frac{z^j}{j!}, \tag{54}$$

and

$$\frac{2\mathrm{Li}_k(1-e^{-z})}{e_\mu(z)+1}e_\mu^{\eta}(z)\sin_\mu^\xi(z) = \sum_{j=0}^\infty \left(\frac{G_{j,\mu}^{(k)}(\eta+i\xi) - G_{j,\mu}^{(k)}(\eta-i\xi)}{2i} \right)\frac{z^j}{j!}. \tag{55}$$

Definition 2. *The degenerate cosine-poly-Genocchi polynomials* $G_{j,\mu}^{(k,c)}(\eta,\xi)$ *and degenerate sine-poly-Genocchi polynomials* $G_{j,\mu}^{(k,s)}(\eta,\xi)$ *for nonnegative integer j are defined, respectively, by*

$$\frac{2\mathrm{Li}_k(1-e^{-z})}{e_\mu(z)+1}e_\mu^{\eta}(z)\cos_\mu^\xi(z) = \sum_{j=0}^\infty G_{j,\mu}^{(k,c)}(\eta,\xi)\frac{z^j}{j!}, \tag{56}$$

and

$$\frac{2\mathrm{Li}_k(1-e^{-z})}{e_\mu(z)+1}e_\mu^{\eta}(z)\sin_\mu^\xi(z) = \sum_{j=0}^\infty G_{j,\mu}^{(k,s)}(\eta,\xi)\frac{z^j}{j!}. \tag{57}$$

On setting $\eta = \xi = 0$ in Equations (56) and (57), we get

$$G_{j,\mu}^{(k,c)}(0,0) = G_{j,\mu}^{(k)}, \quad G_{j,\mu}^{(k,s)}(0,0) = 0, (j \geq 0).$$

Note that $\lim_{\mu\to 0} G_{j,\mu}^{(k,c)}(\eta,\xi) = G_j^{(k,c)}(\eta,\xi)$, $\lim_{\mu\to 0} G_{j,\mu}^{(k,s)}(\eta,\xi) = G_j^{(k,s)}(\eta,\xi)$, $(j \geq 0)$, *where* $G_n^{(k,c)}(\eta,\xi)$ *and* $G_j^{(k,s)}(\eta,\xi)$ *are the new type of poly-Genocchi polynomials.*
From Equations (54)–(57), we determine

$$G_{j,\mu}^{(k,c)}(\eta,\xi) = \frac{G_{j,\mu}^{(k)}(\eta+i\xi) + G_{j,\mu}^{(k)}(\eta-i\xi)}{2} \tag{58}$$

and

$$G_{j,\mu}^{(k,s)}(\eta,\xi) = \frac{G_{j,\mu}^{(k)}(\eta+i\xi) - G_{j,\mu}^{(k)}(\eta-i\xi)}{2i}. \tag{59}$$

Theorem 8. *For $k \in \mathbb{Z}$ and $j \geq 0$, we have*

$$G_{j,\mu}^{(k)}(\eta+i\xi) = \sum_{q=0}^j \binom{j}{q} G_{j-q,\mu}^{(k)}(\eta)(i\xi)_{q,\mu}$$

$$= \sum_{q=0}^{j} \binom{j}{q} G^{(k)}_{j-q,\mu}(\eta + i\xi)_{q,\mu}, \tag{60}$$

and

$$G^{(k)}_{j,\mu}(\eta - i\xi) = \sum_{q=0}^{j} \binom{j}{q} G^{(k)}_{j-q,\mu}(\eta)(-1)^q (i\xi)_{q,\mu}$$

$$= \sum_{q=0}^{j} \binom{j}{q} G^{(k)}_{j-q,\mu}(\eta - i\xi)_{q,\mu}. \tag{61}$$

Proof. On using Equation (52), we see

$$\sum_{j=0}^{\infty} G^{(k)}_{j,\mu}(\eta + i\xi) \frac{z^j}{j!} = \frac{2\mathrm{Li}_k(1 - e^{-z})}{e_\mu(z) + 1} e_\mu^\eta(z) e_\mu^{i\xi}(z)$$

$$= \left(\sum_{j=0}^{\infty} G^{(k)}_{j,\mu}(\eta) \frac{z^j}{j!} \right) \left(\sum_{q=0}^{\infty} (i\xi)_{q,\mu} \frac{z^q}{q!} \right)$$

$$= \sum_{j=0}^{\infty} \left(\sum_{q=0}^{j} \binom{j}{q} G^{(k)}_{j-q,\mu}(\eta)(i\xi)_{q,\mu} \right) \frac{z^j}{j!}. \tag{62}$$

Similarly, we find

$$\frac{2\mathrm{Li}_k(1 - e^{-z})}{e_\mu(z) + 1} e_\mu^\eta(z) e_\mu^{i\xi}(z) = \left(\sum_{j=0}^{\infty} G^{(k)}_{j,\mu} \frac{z^j}{j!} \right) \left(\sum_{q=0}^{\infty} (\eta + i\xi)_{q,\mu} \frac{z^q}{q!} \right)$$

$$= \sum_{j=0}^{\infty} \left(\sum_{q=0}^{j} \binom{j}{q} G^{(k)}_{j-q,\mu}(\eta + i\xi)_{q,\mu} \right) \frac{z^j}{j!}. \tag{63}$$

By comparing the coefficients of z^j on both sides in Equations (62) and (63), we obtain our desired result, Equation (60). The proof of Equation (61) is similar to Equation (60). $\square$

Theorem 9. *If $k \in \mathbb{Z}$ and $j \geq 0$, then*

$$G^{(k,c)}_{j,\mu}(\eta,\xi) = \sum_{r=0}^{j} \binom{j}{r} G^{(k)}_{r,\mu} C_{j-r,\mu}(\eta,\xi)$$

$$= \sum_{r=0}^{\left[\frac{q}{2}\right]} \sum_{q=2r}^{j} \binom{j}{q} \mu^{q-2r}(-1)^r \xi^{2r} S^{(1)}(q,2r) G^{(k)}_{j-q,\mu}(\xi), \tag{64}$$

and

$$G^{(k,s)}_{j,\mu}(\eta,\xi) = \sum_{r=0}^{j} \binom{j}{r} B^{(k)}_{r,\mu} S_{j-r,\mu}(\eta,\xi)$$

$$= \sum_{r=0}^{\left[\frac{q-1}{2}\right]} \sum_{q=2r+1}^{j} \binom{j}{q} \mu^{q-2r-1}(-1)^r \xi^{2r+1} S^{(1)}(q,2r+1) G^{(k)}_{j-q,\mu}(\eta). \tag{65}$$

Proof. From Equations (56) and (10), we see

$$\sum_{j=0}^{\infty} G^{(k,c)}_{j,\mu}(\eta,\xi) \frac{z^j}{j!} = \frac{2\mathrm{Li}_k(1 - e^{-z})}{e_\mu(z) + 1} e_\mu^\eta(t) \cos_\mu^\xi(z)$$

$$= \left(\sum_{r=0}^{\infty} G_{r,\mu}^{(k)} \frac{z^r}{r!} \right) \left(\sum_{j=0}^{\infty} C_{j,\mu}(\eta, \xi) \frac{z^j}{j!} \right)$$

$$= \sum_{j=0}^{\infty} \left(\sum_{r=0}^{j} \binom{j}{r} G_{r,\mu}^{(k)} C_{j-r,\mu}(\eta, \xi) \right) \frac{z^j}{j!}. \tag{66}$$

Similarly, we find

$$\frac{2\mathrm{Li}_k(1 - e^{-z})}{e_\mu(z) + 1} e_\mu^\eta(z) \cos_\mu^\xi(z) = \sum_{j=0}^{\infty} G_{j,\mu}^{(k)}(\eta) \frac{z^j}{j!} \sum_{q=0}^{\infty} \sum_{r=0}^{\lfloor \frac{q}{2} \rfloor} \mu^{q-2r}(-1)^r \xi^{2r} S^{(1)}(q, 2r) \frac{z^r}{r!}$$

$$= \sum_{j=0}^{\infty} \left(\sum_{l=0}^{j} \sum_{m=0}^{\lfloor \frac{l}{2} \rfloor} \binom{j}{l} \mu^{l-2m}(-1)^m \xi^{2m} S^{(1)}(q, 2r) G_{j-q,\mu}^{(k)}(\eta) \right) \frac{z^j}{j!}$$

$$= \sum_{j=0}^{\infty} \left(\sum_{r=0}^{\lfloor \frac{q}{2} \rfloor} \sum_{q=2r}^{j} \binom{j}{q} \mu^{q-2r}(-1)^r \xi^{2r} S^{(1)}(q, 2r) G_{j-q,\mu}^{(k)}(\eta) \right) \frac{z^j}{j!}. \tag{67}$$

By comparing the coefficients of z^j on both sides of Equations (66) and (67), we easily get our first claimed result, Equation (64). Similarly, we can establish our second needed result, Equation (65). □

Theorem 10. *Let $j \geq 0$. Then, we have*

$$G_{j,\mu}^{(2,c)}(\eta, \xi) = \sum_{r=0}^{j} \binom{j}{r} \frac{r! B_r}{r+1} G_{j-r,\mu}^{(c)}(\eta, \xi), \tag{68}$$

and

$$G_{j,\mu}^{(2,s)}(\eta, \xi) = \sum_{r=0}^{j} \binom{j}{r} \frac{r! B_r}{r+1} G_{j-r,\mu}^{(s)}(\eta, \xi). \tag{69}$$

Proof. By using Equation (56), we determine

$$\sum_{j=0}^{\infty} G_{j,\mu}^{(k,c)}(\eta, \xi) \frac{z^j}{j!} = \frac{2\mathrm{Li}_k(1 - e^{-z})}{e_\mu(z) + 1} e_\mu^\eta(z) \cos_\mu^\xi(z)$$

$$= \frac{2 e_\mu^\eta(z) \cos_\mu^\xi(z)}{e_\mu(z) + 1} \int_0^z \underbrace{\frac{1}{e^u - 1} \int_0^u \frac{1}{e^u - 1} \cdots \frac{1}{e^u - 1} \int_0^u \frac{u}{e^u - 1} du \cdots du}_{(k-1)-\text{times}}. \tag{70}$$

On setting $k = 2$ in Equation (70), we find

$$\sum_{j=0}^{\infty} G_{j,\mu}^{(2,c)}(\eta, \xi) \frac{z^j}{j!} = \frac{2 e_\mu^\eta(z) \cos_\mu^\xi(z)}{e_\mu(z) + 1} \int_0^z \frac{u}{e^u - 1} dz$$

$$= \left(\sum_{r=0}^{\infty} \frac{r! B_r z^r}{(r+1)r!} \right) \frac{2z e_\mu^\eta(z) \cos_\mu^\xi(z)}{e_\mu(z) + 1}$$

$$= \left(\sum_{r=0}^{\infty} \frac{r! B_r z^r}{(r+1)r!} \right) \left(\sum_{j=0}^{\infty} G_{j,\mu}^{(c)}(\eta, \xi) \frac{z^j}{j!} \right).$$

On replacing j by $j - r$ in the above equation, we obtain

$$= \sum_{j=0}^{\infty} \sum_{r=0}^{j} \binom{j}{r} \frac{r! B_r}{r+1} G_{j-r,\mu}^{(c)}(\eta, \xi) \frac{z^j}{j!}.$$

Finally, by equating the coefficients of the like powers of z in the last expression, we get the result, Equation (68). The proof of Equation (69) is similar to Equation (68). $\square$

Theorem 11. *For $k \in \mathbb{Z}$ and $j \geq 0$, we have*

$$G_{j,\mu}^{(k,c)}(\eta, \xi) = \sum_{r=0}^{j} \binom{j}{r} \left(\sum_{q=1}^{r+1} \frac{(-1)^{q+r+1} q! S_2(r+1, q)}{q^k (r+1)} \right) G_{j-r,\mu}^{(c)}(\eta, \xi), \tag{71}$$

and

$$G_{j,\mu}^{(k,s)}(\eta, \xi) = \sum_{r=0}^{j} \binom{j}{r} \left(\sum_{q=1}^{r+1} \frac{(-1)^{q+r+1} q! S_2(r+1, q)}{q^k (r+1)} \right) G_{j-r,\mu}^{(s)}(\eta, \xi). \tag{72}$$

Proof. In view of Equations (56) and (11), we see

$$\sum_{j=0}^{\infty} G_{j,\mu}^{(k,c)}(\eta, \xi) \frac{z^j}{j!} = \left(\frac{2 \mathrm{Li}_k(1 - e^{-z})}{z} \right) \left(\frac{z e_\mu^\eta(z) \cos_\mu^\xi(z)}{e_\mu(z) + 1} \right). \tag{73}$$

Now

$$\frac{1}{z} \mathrm{Li}_k(1 - e^{-z}) = \frac{1}{z} \sum_{q=1}^{\infty} \frac{(1 - e^{-z})^q}{q^k}$$

$$= \frac{1}{z} \sum_{q=1}^{\infty} \frac{(-1)^q}{q^k} q! \sum_{r=l}^{\infty} (-1)^r S_2(r, q) \frac{z^r}{r!}$$

$$= \frac{1}{z} \sum_{r=1}^{\infty} \sum_{q=1}^{r} \frac{(-1)^{q+r}}{q^k} q! S_2(r, q) \frac{t^r}{r!}$$

$$= \sum_{r=0}^{\infty} \left(\sum_{q=1}^{r+1} \frac{(-1)^{q+r+1}}{q^k} q! \frac{S_2(r+1, q)}{q+1} \right) \frac{z^r}{r!}. \tag{74}$$

Using Equation (74) in (73), we find

$$\sum_{j=0}^{\infty} G_{j,\mu}^{(k,c)}(\eta, \xi) \frac{z^j}{j!} = \sum_{r=0}^{\infty} \left(\sum_{q=1}^{r+1} \frac{(-1)^{q+r+1}}{q^k} q! \frac{S_2(r+1, q)}{r+1} \right) \frac{z^r}{r!} \left(\sum_{j=0}^{\infty} G_{j,\mu}^{(c)}(\eta, \xi) \frac{z^j}{j!} \right),$$

which on comparing the coefficients of z^j on both sides, yields our desired result, Equation (71). Similarly, we can derive our second result, Equation (72). $\square$

Theorem 12. *Let $k \in \mathbb{Z}$ and $j \geq 0$, then we have*

$$\frac{1}{2} \left[G_{j,\mu}^{(k,c)}(\eta + 1, \xi) + G_{j,\mu}^{(k,c)}(\eta, \xi) \right]$$

$$= \sum_{r=1}^{j} \binom{j}{r} \left(\sum_{q=0}^{r-1} \frac{(-1)^{q+r+1}}{(q+1)^k} (q+1)! S_2(r, q+1) \right) C_{j-r,\mu}(\eta, \xi), \tag{75}$$

and

$$\frac{1}{2} \left[G_{j,\mu}^{(k,s)}(\eta + 1, \xi) + G_{j,\mu}^{(k,s)}(\eta, \xi) \right]$$

$$= \sum_{r=1}^{j} \binom{j}{r} \left(\sum_{q=0}^{r-1} \frac{(-1)^{q+r+1}}{(q+1)^k} (q+1)! S_2(r, q+1) \right) S_{j-r,\mu}(\eta, \xi). \tag{76}$$

Proof. Taking

$$\sum_{j=0}^{\infty} G_{j,\mu}^{(k,c)}(\eta+1, \xi) \frac{z^j}{j!} + \sum_{j=0}^{\infty} G_{j,\mu}^{(k,c)}(\eta, \xi) \frac{z^j}{j!}$$

$$= \frac{2\mathrm{Li}_k(1-e^{-z})}{e_\mu(z)+1} e_\mu^{(\eta+1)}(z) \cos_\mu^{\xi}(z) + \frac{2\mathrm{Li}_k(1-e^{-z})}{e_\mu(z)+1} e_\mu^{(\eta)}(z) \cos_\mu^{\xi}(z)$$

$$= 2\mathrm{Li}_k(1-e^{-z}) e_\mu^{(\eta)}(z) \cos_\mu^{\xi}(z)$$

$$= \sum_{q=0}^{\infty} \frac{(1-e^{-z})^{q+1}}{(q+1)^k} 2 e_\mu^{\eta}(z) \cos_\mu^{(\xi)}(z)$$

$$= \sum_{r=1}^{\infty} \left(\sum_{q=0}^{r-1} \frac{(-1)^{q+r+1}}{(q+1)^k} (q+1)! S_2(r, q+1) \right) \frac{z^r}{r!} 2 e_\mu^{x}(z) \cos_\mu^{(\xi)}(z)$$

$$= 2 \left(\sum_{r=1}^{\infty} \left(\sum_{q=0}^{r-1} \frac{(-1)^{q+r+1}}{(q+1)^k} (q+1)! S_2(r, q+1) \right) \frac{z^r}{r!} \right) \left(\sum_{j=0}^{\infty} C_{j,\mu}(\eta, \xi) \frac{z^j}{j!} \right).$$

On replacing j by $j-r$ in the right side of the above equation, and after comparing the coefficients of z^j on both sides, we acquire the desired result, Equation (75). Similarly, we can obtain the result, Equation (76). $\square$

Theorem 13. *For $k \in \mathbb{Z}$ and $j \geq 0$, we have*

$$G_{j,\mu}^{(k,c)}(\eta+\alpha, \xi) = \sum_{m=0}^{j} \binom{j}{m} G_{j-m,\mu}^{(k,c)}(\eta, \xi)(\alpha)_{m,\mu}, \tag{77}$$

and

$$G_{j,\mu}^{(k,s)}(\eta+\alpha, \xi) = \sum_{m=0}^{j} \binom{j}{m} G_{j-m,\mu}^{(k,s)}(\eta, \xi)(\alpha)_{m,\mu}. \tag{78}$$

Proof. By using Equation (56), we have

$$\sum_{j=0}^{\infty} G_{j,\mu}^{(k,c)}(\eta+\alpha, \xi) \frac{z^j}{j!} = \frac{2\mathrm{Li}_k(1-e^{-z})}{e_\mu(z)+1} e_\mu^{(\eta+\alpha)}(z) \cos_\mu^{(\xi)}(z)$$

$$= \left(\sum_{j=0}^{\infty} G_{j,\mu}^{(k,c)}(\eta, \xi) \frac{z^j}{j!} \right) \left(\sum_{m=0}^{\infty} (\alpha)_{m,\mu} \frac{z^m}{m!} \right)$$

$$= \sum_{j=0}^{\infty} \left(\sum_{m=0}^{j} \binom{j}{m} G_{j-m,\mu}^{(k,c)}(\eta, \xi)(\alpha)_{m,\mu} \right) \frac{z^j}{j!}.$$

By comparing the coefficients of z^j on both sides in the last expression, we acquire our desired result, Equation (77). Similarly, we can derive our second result, Equation (78). $\square$

Theorem 14. *If $k \in \mathbb{Z}$ and $j \geq 0$, then*

$$G_{j,\mu}^{(k,c)}(\eta, \xi) = \sum_{r=0}^{j} \sum_{q=0}^{r} \binom{j}{r} (\eta)_l S_\mu^{(2)}(r, q) G_{j-r,\mu}^{(k,c)}(0, \xi), \tag{79}$$

and

$$G_{j,\mu}^{(k,s)}(\eta,\xi) = \sum_{r=0}^{j}\sum_{q=0}^{r} \binom{j}{r} (\eta)_l S_\mu^{(2)}(r,q) G_{j-r,\mu}^{(k,s)}(0,\xi). \tag{80}$$

Proof. From Equations (56) and (12), we have

$$\sum_{j=0}^{\infty} G_{j,\mu}^{(k,c)}(\eta,\xi)\frac{z^j}{j!} = \frac{2\mathrm{Li}_k(1-e^{-z})}{e_\mu(z)+1}(e_\mu(z)-1+1)^\eta \cos_\mu^\xi(z)$$

$$= \frac{2\mathrm{Li}_k(1-e^{-z})}{e_\mu(z)+1}\sum_{q=0}^{\infty}\binom{\eta}{q}(e_\mu(z)-1)^q \cos_\mu^\xi(z)$$

$$= \frac{2\mathrm{Li}_k(1-e^{-z})}{e_\mu(z)+1}\cos_\mu^\xi(z)\sum_{q=0}^{\infty}(\eta)_q\sum_{r=q}^{\infty}S_\mu^{(2)}(r,q)\frac{z^r}{r!}$$

$$= \sum_{j=0}^{\infty}G_{j,\mu}^{(k,c)}(0,\xi)\frac{z^j}{j!}\sum_{r=0}^{\infty}\left(\sum_{q=0}^{r}(\eta)_q\, S_\mu^{(2)}(r,q)\right)\frac{z^r}{r!}$$

$$= \sum_{j=0}^{\infty}\left(\sum_{r=0}^{j}\sum_{q=0}^{r}\binom{j}{r}(\eta)_q S_\mu^{(2)}(r,q) G_{j-r,\mu}^{(k,c)}(0,\xi)\right)\frac{z^j}{j!}.$$

Finally, by comparing the coefficients of z^j on both sides in the last expression, we arrive at our claimed result, Equation (79). Similarly, we can establish our second result, Equation (80). $\square$

4. Conclusions

In the present article, we have considered the parametric kinds of degenerate poly-Bernoulli and poly-Genocchi polynomials by making use of the degenerate type exponential as well as trigonometric functions. We have also derived some analytical properties of our newly introduced parametric polynomials by using the series manipulation technique. Furthermore, it is noticed that, if we consider any Appell polynomials of a complex variable (as discussed in the present article), then we can easily define its parametric kinds by separating the complex variable into real and imaginary parts.

Author Contributions: All authors contributed equally to the manuscript and typed, read, and approved the final manuscript. All authors have read and agreed to the published version of the manuscript.

Conflicts of Interest: The authors declare no conflict of interest.

Abbreviations

The following abbreviations are used in this manuscript:

MKdV   modified Korteweg–de Vries equation

References

1. Avram, F.; Taqqu, M.S. Noncentral limit theorems and Appell polynomials. *Ann. Probab.* **1987**, *15*, 767–775. [CrossRef]
2. Kim, D.S.; Kim, T.; Lee, H. A note on degenerate Euler and Bernoulli polynomials of complex variable. *Symmetry* **2019**, *11*, 1168; doi:10.3390/sym11091168. [CrossRef]
3. Carlitz, L. Degenerate Stirling Bernoulli and Eulerian numbers. *Util. Math.* **1979**, *15*, 51–88.
4. Carlitz, L. A degenerate Staud-Clausen theorem. *Arch. Math.* **1956**, *7*, 28–33. [CrossRef]
5. Haroon, H.; Khan, W.A. Degenerate Bernoulli numbers and polynomials associated with degenerate Hermite polynomials. *Commun. Korean Math. Soc.* **2018**, *33*, 651–669.

6. Masjed-Jamei, M.; Beyki, M.R.; Koepf, W. A new type of Euler polynomials and numbers. *Mediterr. J. Math.* **2018**, *15*, 138. [CrossRef]

7. Lim, D. Some identities of degenerate Genocchi polynomials. *Bull. Korean Math. Soc.* **2016**, *53*,569–579. [CrossRef]

8. Khan, W.A. A note on Hermite-based poly-Euler and multi poly-Euler polynomials. *Palest. J. Math.* **2017**, *6*, 204–214.

9. Khan, W.A. A note on degenerate Hermite poly-Bernoulli numbers and polynomials. *J. Class. Anal.* **2016**, *8*, 65–76. [CrossRef]

10. Kim, D. A note on the degenerate type of complex Appell polynomials. *Symmetry* **2019**, *11*, 1339. [CrossRef]

11. Ryoo, C.S.; Khan, W.A. On two bivariate kinds of poly-Bernoulli and poly-Genocchi polynomials. *Mathematics* **2020**, *8*, 417. [CrossRef]

12. Sharma, S.K. A note on degenerate poly-Genocchi polynomials. *Int. J. Adv. Appl. Sci.* **2020**, *7*, 1–5. [CrossRef]

13. Kim, D.S.; Kim, T. A note on degenerate poly-Bernoulli numbers polynomials. *Adv. Diff. Equat.* **2015**, *2015*, 258. [CrossRef]

14. Sharma, S.K.; Khan, W.A.; Ryoo, C.S. A parametric kind of the degenerate Fubini numbers and polynomials. *Mathematics* **2020**, *8*, 405. [CrossRef]

15. Kim, T.; Jang, Y.S.; Seo, J.J. A note on poly-Genocchi numbers and polynomials. *Appl. Math. Sci.* **2014**, *8*, 4475–4781. [CrossRef]

16. Kim, T.; Jang, G.-W. A note on degenerate gamma function and degenerate Stirling numbers of the second kind. *Adv. Stud. Contemp. Math.* **2018**, *28*, 207–214.

17. Kim, T. A note on degenerate Stirling polynomials of the second kind. *Proc. Jangjeon Math. Soc.* **2017**, *20*, 319–331.

18. Kim, T.; Yao, Y.; Kim, D.S.; Jang, G.-W. Degenerate r-Stirling numbers and r-Bell polynomials. *Russ. J. Math. Phys.* **2018**, *25*, 44–58. [CrossRef]

19. Kim, T.; Ryoo, C.S. Some identities for Euler and Bernoulli polynomials and their zeros. *Axioms* **2018**, *7*, 56. [CrossRef]

20. Kim, T.; Kim, D. S.; Kim, H. Y.; Jang, L.-C. Degenerate poly-Bernoulli number and polynomials. *Informatica* **2020**, *31*, 2–8.

21. Kim, T.; Kim, D.S.; Kwon, H.-I. A note on degenerate Stirling numbers and their applications. *Proc. Jangjeon Math. Soc.* **2018**, *21*, 195–203.

22. Kim, D. A class of Sheffer sequences of some complex polynomials and their degenerate types. *Mathematics* **2019**, *7*, 1064. [CrossRef]

23. Masjed-Jamei, M.; Beyki, M.R.; Koepf, W. An extension of the Euler-Maclaurin quadrature formula using a parametric type of Bernoulli polynomials. *Bull. Sci. Math.* **2019**, *156*, 102798. [CrossRef]

24. Masjed-Jamei, M.; Koepf, W. Symbolic computation of some power trigonometric series. *J. Symb. Comput.* **2017**, *80*, 273–284. [CrossRef]

25. Masjed-Jamei, M.; Beyki, M.R.; Omey, E. On a parametric kind of Genocchi polynomials. *J. Inq. Spec. Funct.* **2018**, *9*, 68–81.

Article

Some Identities on Type 2 Degenerate Bernoulli Polynomials of the Second Kind

Taekyun Kim [1,2], Lee-Chae Jang [3,*], Dae San Kim [4] and Han Young Kim [2]

[1] School of Sciences, Xian Technological University, Xi'an 710021, China; tkkim@kw.ac.kr
[2] Department of Mathematics, Kwangwoon University, Seoul 139-701, Korea; gksdud213@kw.ac.kr
[3] Graduate School of Education, Konkuk University, Seoul 05029, Korea
[4] Department of Mathematics, Sogang University, Seoul 121-742, Korea; dskim@sogang.ac.kr
* Correspondence: Lcjang@konkuk.ac.kr

Received: 9 March 2020; Accepted: 23 March 2020; Published: 2 April 2020

Abstract: In recent years, many mathematicians studied various degenerate versions of some special polynomials for which quite a few interesting results were discovered. In this paper, we introduce the type 2 degenerate Bernoulli polynomials of the second kind and their higher-order analogues, and study some identities and expressions for these polynomials. Specifically, we obtain a relation between the type 2 degenerate Bernoulli polynomials of the second and the degenerate Bernoulli polynomials of the second, an identity involving higher-order analogues of those polynomials and the degenerate Stirling numbers of the second kind, and an expression of higher-order analogues of those polynomials in terms of the higher-order type 2 degenerate Bernoulli polynomials and the degenerate Stirling numbers of the first kind.

Keywords: type 2 degenerate Bernoulli polynomials of the second kind; degenerate central factorial numbers of the second kind

1. Introduction

In [1,2], Carlitz initiated study of the degenerate Bernoulli and Euler polynomials and obtained some arithmetic and combinatorial results on them. In recent years, many mathematicians have drawn their attention to various degenerate versions of some old and new polynomials and numbers, namely some degenerate versions of Bernoulli numbers and polynomials of the second kind, Changhee numbers of the second kind, Daehee numbers of the second kind, Bernstein polynomials, central Bell numbers and polynomials, central factorial numbers of the second kind, Cauchy numbers, Eulerian numbers and polynomials, Fubini polynomials, Stirling numbers of the first kind, Stirling polynomials of the second kind, central complete Bell polynomials, Bell numbers and polynomials, type 2 Bernoulli numbers and polynomials, type 2 Bernoulli polynomials of the second kind, poly-Bernoulli numbers and polynomials, poly-Cauchy polynomials, and of Frobenius–Euler polynomials, to name a few [3–10] and the references therein.

They have studied those polynomials and numbers with their interest not only in combinatorial and arithmetic properties but also in differential equations and certain symmetric identities [7,9] and references therein, and found many interesting results related to them [3–6,8,10]. It is remarkable that studying degenerate versions is not only limited to polynomials but also extended to transcendental functions. Indeed, the degenerate gamma functions were introduced in connection with degenerate Laplace transforms [11,12].

The motivation for this research is to introduce the type 2 degenerate Bernoulli polynomials of the second kind defined by

$$\frac{(1+t) - (1+t)^{-1}}{\log_\lambda(1+t)}(1+t)^x = \sum_{n=0}^{\infty} b_{n,\lambda}^*(x)\frac{t^n}{n!},$$

and investigate its arithmetic and combinatorial properties. The facts in Section 1 are some known definitions and results that are needed throughout this paper. However, all of the results in Section 2 are new.

We will spend the rest of this section in recalling some necessary stuffs for the next section.

As is known, the type 2 Bernoulli polynomials are defined by the generating function [5,13]

$$\frac{t}{e^t - e^{-t}}e^{xt} = \sum_{n=0}^{\infty} B_n^*(x)\frac{t^n}{n!}. \tag{1}$$

From (1), we note that

$$B_n^*(x) = 2^{n-1}B_n\left(\frac{x+1}{2}\right), \quad (n \geq 0), \tag{2}$$

where $B_n(x)$ are the ordinary Bernoulli polynomials given by

$$\frac{t}{e^t - 1}e^{xt} = \sum_{n=0}^{\infty} B_n(x)\frac{t^n}{n!}.$$

Also, the type 2 Euler polynomials are given by [5,13]

$$e^{xt}\mathrm{sech}t = \frac{2}{e^t + e^{-t}}e^{xt} = \sum_{n=0}^{\infty} E_n^*(x)\frac{t^n}{n!}. \tag{3}$$

Note that

$$E_n^*(x) = 2^n E_n\left(\frac{x+1}{2}\right), \quad (n \geq 0), \tag{4}$$

where $E_n(x)$ are the ordinary Euler polynomials given by [14,15]

$$\frac{2}{e^t + 1}e^{xt} = \sum_{n=0}^{\infty} E_n(x)\frac{t^n}{n!}.$$

The central factorial numbers of the second kind are defined as [5,8]

$$x^n = \sum_{k=0}^{n} T(n,k)x^{[k]}, \tag{5}$$

or equivalently as

$$\frac{1}{k!}\left(e^{\frac{t}{2}} - e^{-\frac{t}{2}}\right)^k = \sum_{n=k}^{\infty} T(n,k)\frac{t^n}{n!}, \tag{6}$$

where $x^{[0]} = 1$, $x^{[n]} = x\left(x + \frac{n}{2} - 1\right)\left(x + \frac{n}{2} - 2\right)\cdots\left(x - \frac{n}{2} + 1\right)$, $(n \geq 1)$.

It is well known that the Daehee polynomials are defined by [16,17]

$$\frac{\log(1+t)}{t}(1+t)^x = \sum_{k=0}^{n} D_n(x)\frac{t^n}{n!}. \tag{7}$$

When $x = 0$, $D_n = D_n(0)$ are called the Daehee numbers.

The Bernoulli polynomials of the second kind of order r are defined by [15]

$$\left(\frac{t}{\log(1+t)}\right)^r (1+t)^x = \sum_{k=0}^{n} b_n^{(r)}(x)\frac{t^n}{n!}. \tag{8}$$

Note that $b_n^{(r)}(x) = B_n^{(n-r+1)}(x+1)$, $(n \geq 0)$. Here $B_n^{(r)}(x)$ are the ordinary Bernoulli polynomials of order r given by [8,15–18]

$$\left(\frac{t}{e^t-1}\right)^r e^{xt} = \sum_{k=0}^{n} B_n^{(r)}(x)\frac{t^n}{n!}. \tag{9}$$

It is known that the Stirling numbers of the second kind are defined by [8]

$$\frac{1}{k!}\left(e^t-1\right)^k = \sum_{n=k}^{\infty} S_2(n,k)\frac{t^n}{n!}, \tag{10}$$

and the Stirling numbers of the first kind by [8]

$$\frac{1}{k!}\log^k(1+t) = \sum_{n=k}^{\infty} S_1(n,k)\frac{t^n}{n!}. \tag{11}$$

For any nonzero $\lambda \in \mathbb{R}$, the degenerate exponential function is defined by [11,12]

$$e_\lambda^x(t) = (1+\lambda t)^{\frac{x}{\lambda}} = \sum_{n=0}^{\infty} (x)_{n,\lambda}\frac{t^n}{n!}, \tag{12}$$

where $(x)_{0,\lambda} = 1$, $(x)_{n,\lambda} = x(x-\lambda)\cdots(x-(n-1)\lambda)$, $(n \geq 1)$.

In particular, we let

$$e_\lambda(t) = e_\lambda^1(t) = (1+\lambda t)^{\frac{1}{\lambda}}. \tag{13}$$

In [1,2], Carlitz introduced the degenerate Bernoulli polynomials which are given by the generating function

$$\frac{t}{e_\lambda(t)-1}e_\lambda^x(t) = \sum_{n=0}^{\infty} \beta_{n,\lambda}(x)\frac{t^n}{n!}. \tag{14}$$

Also, he considered the degenerate Euler polynomials given by [1,2]

$$\frac{2}{e_\lambda(t)+1}e_\lambda^x(t) = \sum_{n=0}^{\infty} \mathcal{E}_{n,\lambda}(x)\frac{t^n}{n!}. \tag{15}$$

Recently, Kim-Kim considered the degenerate central factorial numbers of the second kind given by [8,13]

$$\frac{1}{k!}\left(e_\lambda^{\frac{1}{2}}(t) - e_\lambda^{-\frac{1}{2}}(t)\right)^k = \sum_{n=k}^{\infty} T_\lambda(n,k)\frac{t^n}{n!}. \tag{16}$$

Note that $\lim_{\lambda \to 0} T_\lambda(n,k) = T(n,k)$.

2. Type 2 Degenerate Bernoulli Polynomials of the Second Kind

Let $\log_\lambda t$ be the compositional inverse of $e_\lambda(t)$ in (13). Then we have

$$\log_\lambda t = \frac{1}{\lambda}\left(t^\lambda - 1\right). \tag{17}$$

Note that $\lim_{\lambda \to 0} \log_\lambda t = \log t$. Now, we define the *degenerate Daehee polynomials* by

$$\frac{\log_\lambda(1+t)}{t}(1+t)^x = \sum_{n=0}^{\infty} D_{n,\lambda}(x)\frac{t^n}{n!}. \tag{18}$$

Note that $\lim_{\lambda \to 0} D_{n,\lambda}(x) = D_n(x)$, $(n \geq 0)$. In view of (8), we also consider the degenerate Bernoulli polynomials of the second kind of order α given by

$$\left(\frac{t}{\log_\lambda(1+t)}\right)^{\alpha}(1+t)^x = \sum_{n=0}^{\infty} b_{n,\lambda}^{(\alpha)}(x)\frac{t^n}{n!}. \tag{19}$$

Note that $\lim_{\lambda \to 0} b_{n,\lambda}^{(\alpha)}(x) = b_n^{(\alpha)}(x)$, $(n \geq 0)$. From (19), we have

$$\left(\frac{\lambda t}{(1+t)^{\frac{\lambda}{2}} - (1+t)^{-\frac{\lambda}{2}}}\right)^{\alpha}(1+t)^{x-\frac{\lambda\alpha}{2}} = \sum_{n=0}^{\infty} b_{n,\lambda}^{(\alpha)}(x)\frac{t^n}{n!}. \tag{20}$$

For $\alpha = r \in \mathbb{N}$, and replacing t by $e^{2t} - 1$ in (20), we get

$$\sum_{m=0}^{\infty} b_{m,\lambda}^{(r)}(x)\frac{1}{m!}(e^{2t} - 1)^m = \left(\frac{\lambda t}{e^{t\lambda} - e^{-t\lambda}}\right)^r \frac{1}{t^r}(e^{2t} - 1)^r e^{(2x - \lambda r)t}$$

$$= \sum_{k=0}^{\infty} B_k^*\left(\frac{2x}{\lambda} - r\right)\frac{\lambda^k t^k}{k!}\sum_{m=0}^{\infty} S_2(m+r,r)2^{m+r}\frac{1}{\binom{m+r}{r}}\frac{t^m}{m!}$$

$$= \sum_{n=0}^{\infty}\left(\sum_{m=0}^{n}\binom{n}{m}B_{n-m}^*\left(\frac{2x}{\lambda} - r\right)\lambda^{n-m}\frac{S_2(m+r,r)}{\binom{m+r}{r}}2^{m+r}\right)\frac{t^n}{n!}. \tag{21}$$

On the other hand,

$$\sum_{m=0}^{\infty} b_{m,\lambda}^{(r)}(x)\frac{1}{m!}(e^{2t} - 1)^m = \sum_{m=0}^{\infty} b_{m,\lambda}^{(r)}(x)\sum_{n=m}^{\infty} S_2(n,m)2^n\frac{t^n}{n!}$$

$$= \sum_{n=0}^{\infty}\left(\sum_{m=0}^{n} b_{m,\lambda}^{(r)}(x)2^n S_2(n,m)\right)\frac{t^n}{n!}. \tag{22}$$

From (21) and (22), we have

$$\sum_{m=0}^{n} b_{m,\lambda}^{(r)}(x)S_2(n,m) = \sum_{m=0}^{n}\binom{n}{m}B_{n-m}^*\left(\frac{2x}{\lambda} - r\right)\lambda^{n-m}\frac{S_2(m+r,r)}{\binom{m+r}{r}}2^{m+r-n}. \tag{23}$$

Now, we define the *type 2 degenerate Bernoulli polynomials of the second kind* by

$$\frac{(1+t) - (1+t)^{-1}}{\log_\lambda(1+t)}(1+t)^x = \sum_{n=0}^{\infty} b_{n,\lambda}^*(x)\frac{t^n}{n!}. \tag{24}$$

When $x = 0$, $b_{n,\lambda}^* = b_{n,\lambda}^*(0)$ are called the type 2 degenerate Bernoulli numbers of the second kind. Note that $\lim_{\lambda \to 0} b_{n,\lambda}^*(x) = b_n^*(x)$, where $b_n^*(x)$ are the type 2 Bernoulli polynomials of the second kind given by

$$\frac{(1+t) - (1+t)^{-1}}{\log(1+t)}(1+t)^x = \sum_{n=0}^{\infty} b_n^*(x)\frac{t^n}{n!}.$$

From (19) and (24), we note that

$$
\frac{(1+t)-(1+t)^{-1}}{\log_\lambda(1+t)}(1+t)^x = \frac{t}{\log_\lambda(1+t)}(1+t)^x\left(1+\frac{1}{1+t}\right)
$$

$$
= \frac{t}{\log_\lambda(1+t)}(1+t)^x + \frac{t}{\log_\lambda(1+t)}(1+t)^{x-1}
$$

$$
= \sum_{n=0}^{\infty}\left(b_{n,\lambda}^{(1)}(x)+b_{n,\lambda}^{(1)}(x-1)\right)\frac{t^n}{n!}. \tag{25}
$$

Therefore, we obtain the following theorem.

Theorem 1. *For $n \geq 0$, we have*

$$
b_{n,\lambda}^*(x) = b_{n,\lambda}^{(1)}(x) + b_{n,\lambda}^{(1)}(x-1).
$$

Moreover,

$$
\sum_{m=0}^{n} b_{m,\lambda}^{(r)}(x)S_2(n,m) = \sum_{m=0}^{n}\binom{n}{m}B_{n-m}^*\left(\frac{2x}{\lambda}-r\right)\lambda^{n-m}\frac{S_2(m+r,r)}{\binom{m+r}{r}}2^{m+r-n},
$$

where r is a positive integer.

Now, we observe that

$$
\frac{(1+t)-(1+t)^{-1}}{\log_\lambda(1+t)}(1+t)^x = \sum_{l=0}^{\infty}b_{l,\lambda}^*\frac{t^l}{l!}\sum_{m=0}^{\infty}(x)_m\frac{t^m}{m!}
$$

$$
= \sum_{n=0}^{\infty}\left(\sum_{l=0}^{n}\binom{n}{l}b_{l,\lambda}^*(x)_{n-l}\right)\frac{t^n}{n!}, \tag{26}
$$

where $(x)_0 = 1$, $(x)_n = x(x-1)\cdots(x-n+1)$, $(n \geq 1)$. From (24) and (26), we get

$$
b_{n,\lambda}^*(x) = \sum_{l=0}^{n}\binom{n}{l}b_{l,\lambda}^*(x)_{n-l}, \quad (n \geq 0). \tag{27}
$$

For $\alpha \in \mathbb{R}$, let us define the *type 2 degenerate Bernoulli polynomials of the second kind of order α* by

$$
\left(\frac{(1+t)-(1+t)^{-1}}{\log_\lambda(1+t)}\right)^\alpha(1+t)^x = \sum_{n=0}^{\infty}b_{n,\lambda}^{*(\alpha)}(x)\frac{t^n}{n!} \tag{28}
$$

When $x = 0$, $b_{n,\lambda}^{*(\alpha)} = b_{n,\lambda}^{*(\alpha)}(0)$ are called the type 2 degenerate Bernoulli numbers of the second kind of order α.

Let $\alpha = k \in \mathbb{N}$. Then we have

$$
\sum_{n=0}^{\infty}b_{n,\lambda}^{*(k)}(x)\frac{t^n}{n!} = \left(\frac{(1+t)-(1+t)^{-1}}{\log_\lambda(1+t)}\right)^k(1+t)^x. \tag{29}
$$

By replacing t by $e_\lambda(t) - 1$ in (29), we get

$$
\begin{aligned}
\frac{k!}{t^k}\frac{1}{k!}\left(e_\lambda(t) - e_\lambda^{-1}(t)\right)^k e_\lambda^x(t) &= \sum_{l=0}^{\infty} b_{l,\lambda}^{*(k)}(x)\frac{1}{l!}(e_\lambda(t) - 1)^l \\
&= \sum_{l=0}^{\infty} b_{l,\lambda}^{*(k)}(x)\sum_{n=l}^{\infty} S_{2,\lambda}(n,l)\frac{t^n}{n!} \\
&= \sum_{n=0}^{\infty}\left(\sum_{l=0}^{n} b_{l,\lambda}^{*(k)}(x)S_{2,\lambda}(n,l)\right)\frac{t^n}{n!},
\end{aligned}
\tag{30}
$$

where $S_{2,\lambda}(n,l)$ are the degenerate Stirling numbers of the second kind given by [6]

$$
\frac{1}{k!}(e_\lambda(t) - 1)^k = \sum_{n=k}^{\infty} S_{2,\lambda}(n,k)\frac{t^n}{n!}.
\tag{31}
$$

On the other hand, we also have

$$
\begin{aligned}
\frac{k!}{t^k}\frac{1}{k!}\left(e_\lambda(t) - e_\lambda^{-1}(t)\right)^k e_\lambda^x(t) &= \frac{k!}{t^k}\frac{1}{k!}\left(e_\lambda^2(t) - 1\right)^k e_\lambda^{x-k}(t) \\
&= \frac{k!}{t^k}\frac{1}{k!}\left(e_{\frac{\lambda}{2}}(2t) - 1\right)^k e_\lambda^{x-k}(t) \\
&= \sum_{m=0}^{\infty} S_{2,\frac{\lambda}{2}}(m+k,k)\frac{2^{m+k}}{\binom{m+k}{k}}\frac{t^m}{m!}\sum_{l=0}^{\infty}(x-k)_{l,\lambda}\frac{t^l}{l!} \\
&= \sum_{n=0}^{\infty}\left(\sum_{m=0}^{n}\frac{\binom{n}{m}2^{m+k}}{\binom{m+k}{k}}S_{2,\frac{\lambda}{2}}(m+k,k)(x-k)_{n-m,\lambda}\right)\frac{t^n}{n!}.
\end{aligned}
\tag{32}
$$

Therefore, by (30) and (32), we obtain the following theorem.

Theorem 2. *For $n \geq 0$, we have*

$$
\sum_{l=0}^{n} b_{l,\lambda}^{*(k)}(x)S_{2,\lambda}(n,l) = \sum_{l=0}^{n}\frac{\binom{n}{l}2^{l+k}}{\binom{l+k}{k}}S_{2,\frac{\lambda}{2}}(l+k,k)(x-k)_{n-l,\lambda}.
$$

In particular,

$$
2^{n+k}S_{2,\frac{\lambda}{2}}(n+k,k) = \binom{n+k}{k}\sum_{l=0}^{n} b_{l,\lambda}^{*(k)}(k)S_{2,\lambda}(n,l).
$$

For $\alpha \in \mathbb{R}$, we recall that the type 2 degenerate Bernoulli polynomials of order α are defined by [5,13]

$$
\left(\frac{t}{e_\lambda(t) - e_\lambda^{-1}(t)}\right)^\alpha e_\lambda^x(t) = \sum_{n=0}^{\infty} \beta_{n,\lambda}^{*(\alpha)}(x)\frac{t^n}{n!}.
\tag{33}
$$

For $k \in \mathbb{N}$, let us take $\alpha = -k$ and replace t by $\log_\lambda(1+t)$ in (33). Then we have

$$
\begin{aligned}
\left(\frac{(1+t) - (1+t)^{-1}}{\log_\lambda(1+t)}\right)^k (1+t)^x &= \sum_{l=0}^{\infty} \beta_{l,\lambda}^{*(-k)}(x)\frac{1}{l!}\left(\log_\lambda(1+t)\right)^l \\
&= \sum_{l=0}^{\infty} \beta_{l,\lambda}^{*(-k)}(x)\sum_{n=l}^{\infty} S_{1,\lambda}(n.l)\frac{t^n}{n!} \\
&= \sum_{n=0}^{\infty}\left(\sum_{l=0}^{n} \beta_{l,\lambda}^{*(-k)}S_{1,\lambda}(n.l)\right)\frac{t^n}{n!},
\end{aligned}
\tag{34}
$$

where $S_{1,\lambda}(n,l)$ are the degenerate Stirling numbers of the first kind given by

$$\frac{1}{k!}(\log_\lambda(1+t))^k = \sum_{n=k}^{\infty} S_{1,\lambda}(n,k)\frac{t^n}{n!}. \tag{35}$$

Note here that $\lim_{\lambda\to 0} S_{1,\lambda}(n,l) = S_1(n,l)$. Therefore, by (26) and (34), we obtain the following theorem.

Theorem 3. *For $n \geq 0$ and $k \in \mathbb{N}$, we have*

$$b_{n,\lambda}^{*(k)}(x) = \sum_{l=0}^{n} \beta_{l,\lambda}^{*(-k)}(x)S_{1,\lambda}(n,l).$$

We observe that

$$\frac{1}{k!}t^k = \frac{1}{k!}\left((1+t)^{\frac{1}{2}} - (1+t)^{-\frac{1}{2}}\right)^k (1+t)^{\frac{k}{2}}$$

$$= \frac{1}{k!}\left(e_\lambda^{\frac{1}{2}}(\log_\lambda(1+t)) - e_\lambda^{-\frac{1}{2}}(\log_\lambda(1+t))\right)^k (1+t)^{\frac{k}{2}}$$

$$= \sum_{l=k}^{\infty} T_\lambda(l,k)\frac{1}{l!}(\log_\lambda(1+t))^l \sum_{r=0}^{\infty} \binom{k}{2}_r \frac{t^r}{r!}$$

$$= \sum_{l=k}^{\infty} T_\lambda(l,k) \sum_{m=l}^{\infty} S_{1,\lambda}(m,l)\frac{t^m}{m!} \sum_{r=0}^{\infty} \binom{k}{2}_r \frac{t^r}{r!}$$

$$= \sum_{m=k}^{\infty} \sum_{l=k}^{m} T_\lambda(l,k)S_{1,\lambda}(m,l)\frac{t^m}{m!} \sum_{r=0}^{\infty} \binom{k}{2}_r \frac{t^r}{r!}$$

$$= \sum_{n=k}^{\infty}\left(\sum_{m=k}^{n}\sum_{l=k}^{m} T_\lambda(l,k)S_{1,\lambda}(m,l)\binom{n}{m}\binom{k}{2}_{n-m}\right)\frac{t^n}{n!}. \tag{36}$$

On the other hand,

$$\frac{1}{k!}t^k = \left(\frac{t}{\log_\lambda(1+t)}\right)^k \frac{1}{k!}(\log_\lambda(1+t))^k$$

$$= \sum_{l=0}^{\infty} b_{l,\lambda}^{(k)}\frac{t^l}{l!} \sum_{m=k}^{\infty} S_{1,\lambda}(m,k)\frac{t^m}{m!}$$

$$= \sum_{n=k}^{\infty}\left(\sum_{m=k}^{n} S_{1,\lambda}(m,k)b_{n-m,\lambda}^{(k)}\binom{n}{m}\right)\frac{t^n}{n!}. \tag{37}$$

Therefore, by (36) and (37), we obtain the following theorem.

Theorem 4. *For $n, k \geq 0$, we have*

$$\sum_{m=k}^{n}\sum_{l=k}^{m} T_\lambda(l,k)S_{1,\lambda}(m,l)\binom{n}{m}\binom{k}{2}_{n-m} = \sum_{m=k}^{n} S_{1,\lambda}(m,k)b_{n-m,\lambda}^{(k)}\binom{n}{m}.$$

3. Conclusions

In this paper, we introduced the type 2 degenerate Bernoulli polynomials of the second kind and their higher-order analogues, and studied some identities and expressions for these polynomials. Specifically, we obtained a relation between the type 2 degenerate Bernoulli polynomials of the second and the degenerate Bernoulli polynomials of the second, an identity involving higher-order analogues of those polynomials and the degenerate Stirling numbers of second kind, and an expression of higher-order analogues of those polynomials in terms of the higher-order type 2 degenerate Bernoulli polynomials and the degenerate Stirling numbers of the first kind.

In addition, we obtained an identity involving the higher-order degenerate Bernoulli polynomials of the second kind, the type 2 Bernoulli polynomials and Stirling numbers of the second kind, and an identity involving the degenerate central factorial numbers of the second kind, the degenerate Stirling numbers of the first kind and the higher-order degenerate Bernoulli polynomials of the second kind.

Next, we would like to mention three possible applications of our results. The first one is their applications to identities of symmetry. For instance, in [7] by using the p-adic fermionic integrals it was possible for us to find many symmetric identities in three variables related to degenerate Euler polynomials and alternating generalized falling factorial sums.

The second one is their applications to differential equations. Indeed, in [9] we derived an infinite family of nonlinear differential equations having the generating function of the degenerate Changhee numbers of the second kind as a solution. As a result, from those differential equations we obtained an interesting identity involving the degenerate Changhee and higher-order degenerate Changhee numbers of the second kind.

The third one is their applications to probability. For example, in [19,20] we showed that both the degenerate λ-Stirling polynomials of the second and the r-truncated degenerate λ-Stirling polynomials of the second kind appear in certain expressions of the probability distributions of appropriate random variables.

These possible applications of our results require a considerable amount of work and they should appear as separate papers. We have witnessed in recent years that studying various degenerate versions of some special polynomials and numbers are very fruitful and promising [21]. It is our plan to continue to do this line of research, as one of our near future projects.

Author Contributions: T.K. and D.S.K. conceived of the framework and structured the whole paper; D.S.K. and T.K. wrote the paper; L.-C.J. and H.Y.K. checked the results of the paper; D.S.K. and T.K. completed the revision of the article. All authors have read and agreed to the published version of the manuscript.

Funding: This research received no external funding.

Conflicts of Interest: The authors declare that they have no competing interests.

References

1. Carlitz, L. Degenerate Stirling, Bernoulli and Eulerian numbers. *Utilitas Math.* **1979**, *15*, 51–88.
2. Carlitz, L. A degenerate Staudt-Clausen theorem. *Arch. Math.* **1956**, *7*, 28–33. [CrossRef]
3. Dolgy, D.V.; Kim, T. Some explicit formulas of degenerate Stirling numbers associated with the degenerate special numbers and polynomials. *Proc. Jangjeon Math. Soc.* **2018**, *21*, 309–317.
4. Haroon, H.; Khan, W.A. Degenerate Bernoulli numbers and polynomials associated with degenerate Hermite polynomials. *Commun. Korean Math. Soc.* **2018**, *33*, 651–669.
5. Jang, G.-W.; Kim, T. A note on type 2 degenerate Euler and Bernoulli polynomials. *Adv. Stud. Contemp. Math.* **2019**, *29*, 147–159.
6. Kim, T. A note on degenerate Stirling polynomials of the second kind. *Proc. Jangjeon Math. Soc.* **2017**, *20*, 319–331.
7. Kim, T.; Kim, D.S. Identities of symmetry for degenerate Euler polynomials and alternating generalized falling factorial sums. *Iran. J. Sci. Technol. Trans. Sci.* **2017**, *41*, 939–949. [CrossRef]
8. Kim, T.; Kim, D.S. Degenerate central factorial numbers of the second kind. *Rev. R. Acad. Cienc. Exactas Fis. Nat. Ser. A Mat. RACSAM* **2019**, 1–9. [CrossRef]
9. Kim, T.; Kim, D.S. Differential equations associated with degenerate Changhee numbers of the second kind. *Rev. R. Acad. Cienc. Exactas Fís. Nat. Ser. A Mat. RACSAM* **2019**, *113*, 1785–1793. [CrossRef]
10. Kim, T.; Yao, Y.; Kim, D.S.; Jang, G.-W. Degenerate r-Stirling numbers and r-Bell polynomials. *Russ. J. Math. Phys.* **2018**, *25*, 44–58. [CrossRef]
11. Kim, T.; Jang, G.-W. A note on degenerate gamma function and degenerate Stirling numbers of the second kind. *Adv. Stud. Contemp. Math.* **2018**, *28*, 207–214.
12. Kim, T.; Kim, D.S. Degenerate Laplace transfrom and degenerate gamma funmction. *Russ. J. Math. Phys.* **2017**, *24*, 241–248. [CrossRef]

13. Kim, T.; Kim, D.S. A note on type 2 Changhee and Daehee polynomials. *Rev. R. Acad. Cienc. Exactas Fis. Nat. Ser. A Mat. RACSAM* **2019**, *113*, 2783–2791. [CrossRef]
14. He, Y.; Araci, S. Sums of products of Apostol-Bernoulli and Apostol-Euler polynomials. *Adv. Differ. Equ.* **2014**, *2014*, 13. [CrossRef]
15. Roman, S. The umbral calculus. In *Pure and Applied Mathematics*; Academic Press Inc.: New York, NY, USA, 1984; p. 193, ISBN 0-12-594380-6.
16. Simsek, Y. Identities and relations related to combinatorial numbers and polynomials. *Proc. Jangjeon Math. Soc.* **2017**, *20*, 127–135.
17. Simsek, Y. Identities on the Changhee numbers and Apostol-type Daehee polynomials. *Adv. Stud. Contemp. Math.* **2011**, *27*, 199–212.
18. Zhang, W.; Lin, X. Identities invoving trigonometric functions and Bernoulli numbers. *Appl. Math. Comput.* **2018**, *334*, 288–294.
19. Kim, T.; Kim, D.S.; Kim, H.Y.; Kwon, J. Degenerate Stirling polynomials of the second kind and some applications. *Symmetry* **2019**, *11*, 1046. [CrossRef]
20. Kim, T.; Kim, D.S. Some identities of extended degenerate r-central Bell polynomials arising from umbral calculus. *Rev. R. Acad. Cienc. Exactas Fís. Nat. Ser. A Mat. RACSAM* **2020**, *114*, 19. [CrossRef]
21. Kim, T.; Kim, D.S.; Kim, H.Y.; Jang, L.-C. Degenerate poly-Bernoulli numbers and polynomials. *Informatica* **2020**, *31*, 2–8.

Article

Exceptional Set for Sums of Symmetric Mixed Powers of Primes

Jinjiang Li [1], Chao Liu [1], Zhuo Zhang [1] and Min Zhang [2,*]

[1] Department of Mathematics, China University of Mining and Technology, Beijing 100083, China;
     jinjiang.li.math@gmail.com (J.L.); chao.liu@student.cumtb.edu.cn (C.L.);
     zhuo.zhang.math@foxmail.com (Z.Z.)
[2] School of Applied Science, Beijing Information Science and Technology University, Beijing 100192, China
* Correspondence: min.zhang.math@gmail.com

Received: 17 January 2020; Accepted: 25 February 2020; Published: 2 March 2020

Abstract: The main purpose of this paper is to use the Hardy–Littlewood method to study the solvability of mixed powers of primes. To be specific, we consider the even integers represented as the sum of one prime, one square of prime, one cube of prime, and one biquadrate of prime. However, this representation can not be realized for all even integers. In this paper, we establish the exceptional set of this kind of representation and give an upper bound estimate.

Keywords: Waring–Goldbach problem; circle method; exceptional set; symmetric form

MSC: 11P05, 11P32, 11P55

---

1. Introduction and Main Result

Let N, k_1, $k_2, \ldots, k_s$ be natural numbers which satisfy $2 \leqslant k_1 \leqslant k_2 \leqslant \cdots \leqslant k_s$, $N > s$. Waring's problem of unlike powers concerns the possibility of representation of N in the form

$$N = x_1^{k_1} + x_2^{k_2} + \cdots + x_s^{k_s}. \tag{1}$$

For previous literature, the reader could refer to section P12 of LeVeque's *Reviews in number theory* and the bibliography of Vaughan [1]. For the special case, $k_1 = k_2 = \cdots = k_s$, an interesting problem is to determine the value for $k \geqslant 2$, called Waring's problem, of the function $G(k)$, the least positive number s such that every *sufficiently large* number can be represented the sum of at most s k-th powers of natural numbers. For this problem, there are only two values of the function $G(k)$ determined exactly. To be specific, $G(2) = 4$, by Lagrange in 1770, and $G(4) = 16$, by Davenport [2]. The majority of information for $G(k)$ has been derived from the Hardy–Littlewood method. This method has arised from a celebrated paper of Hardy and Ramanujan [3], which focused on the partition function.

There are many authors who devoted to establish many kinds of generalisations of this classical version of Waring's problem. Among these results, it is necessary to illustrate some of the majority variants. We begin with the most famous Waring–Goldbach problem, for which one devotes to investigate the possibility of the representation of integers as sums of k-th powers of prime numbers. In order to explain the associated congruence conditions, we denote by k a natural number and p a prime number. We write $\theta = \theta(k; p)$ as the integer with the properties $p^\theta | k$ and $p^\theta \nmid k$, and then define $\gamma = \gamma(k, p)$ by

$$\gamma(k, p) = \begin{cases} \theta + 2, & \text{when } p = 2 \text{ and } \theta > 0, \\ \theta + 1, & \text{otherwise.} \end{cases}$$

Also, we set

$$K(k) = \prod_{(p-1)|k} p^{\gamma}.$$

Denote by $H(k)$ the smallest integer s, which satisfies every sufficiently large integer congruent to s modulo $K(k)$ can be represented as the sum of s k-th powers of primes . By noting the fact that for $(p-1)|k$, we have $p^{\theta}(p-1)|k$, provided that $a^k \equiv 1 \pmod{p^{\gamma}}$ and $(p,a) = 1$. This states the seemingly awkward definition of $H(k)$, because if n is the sum of s k-th powers of primes exceeding $k+1$, then it must satisfy $n \equiv s \pmod{K(k)}$. Trivially, further congruence conditions could arise from the primes p which satisfy $(p-1) \nmid k$. Following the previous investigations of Vinogradov [4,5], Hua systematically considered and investigated the additive problems involving prime variables in his famous book (see Hua [6,7]).

For the nonhomogeneous case, the most optimistic conjecture suggests that, for each prime p, if the Equation (1) has p-adic solutions and satisfies

$$k_1^{-1} + k_2^{-1} + \cdots + k_s^{-1} > 1, \tag{2}$$

then n can be written as the sum of unlike powers of positive integers (1) provided that n is sufficiently large in terms of k. For $s = 3$, such an claim maybe not true in certain situations (see Jagy and Kaplansky [8], or Exercise 5 of Chapter 8 of Vaughan [1]). However, a guide of application for the Hardy–Littlewood method suggests that the condition (2) should ensure at least that *almost all* integers satisfying the expected congruence conditions can be represented. Moreover, once subject to the following condition

$$k_1^{-1} + k_2^{-1} + \cdots + k_s^{-1} > 2, \tag{3}$$

a standard application of the Hardy–Littlewood method suggests that all the integers, which satisfy necessary congruence conditions, could be written in the form (1). Meanwhile, a conventional argument of the circle method shows that in situations in which the condition (2) does not hold, then every sufficiently large integer can not be represented in the expected form.

Since the Hardy–Littlewood method, the investigation of Waring's problem for unlike powers has produced splendid progress in circle method, especially for the classical version of Waring's problem. Additive Waring's problems of unlike powers involving squares, cubes or biquadrates offen attract greater interest of many mathematicians than those cases with higher mixed powers, and the current circumstance is quite satisfactory. For example, the reader can refer to references [9–19].

The Waring–Goldbach problem of mixed powers concerns the representation of N which satisfying some necessary congruence conditions as the form

$$N = p_1^{k_1} + p_2^{k_2} + \cdots + p_s^{k_s},$$

where $p_1, p_2, \ldots, p_s$ are prime variables.

In 2002, Brüdern and Kawada [20] proved that for every sufficiently large even integer N, the equation

$$N = x + p_2^2 + p_3^3 + p_4^4$$

is solvable with x being an almost–prime $\mathcal{P}_2$ and the p_j $(j = 2,3,4)$ primes. As usual, $\mathcal{P}_r$ denotes an almost–prime with at most r prime factors, counted according to multiplicity. On the other hand, in 2015, Zhao [21] established that, for $k = 3$ or 4, every sufficiently large even integer N can be represented as the form

$$N = p_1 + p_2^2 + p_3^3 + p_4^k + 2^{v_1} + 2^{v_2} + \cdots + 2^{v_{t(k)}},$$

where $p_1, \ldots, p_4$ are primes, $v_1, v_2, \ldots, v_{t(k)}$ are natural numbers, and $t(3) = 16$, $t(4) = 18$, which is an improvement result of Liu and Lü [22]. Afterwards, Lü [23] improved the result of Zhao [21] and showed that every sufficiently large even integer N can be represented as a sum of one prime, one square of prime, one cube of prime, one biquadrate of prime and 16 powers of 2.

In view of the results of Brüdern and Kawada [20], Zhao [21], Liu and Lü [22] and Lü [23], it is reasonable to conjecture that, for sufficiently large integer N satisfying $N \equiv 0 \, (\mathrm{mod}\, 2)$, the following Diophantine equation

$$N = p_1 + p_2^2 + p_3^3 + p_4^4$$

is solvable, here and below the letter p, with or without subscript, always denotes a prime number. However, this conjecture may be out of reach at present with the known methods and techniques.

In this paper, we shall consider the exceptional set of the problem (4) and establish the following result.

Theorem 1. *Let $E(N)$ denote the number of positive integers n, which satisfy $n \equiv 0 \,(\mathrm{mod}\, 2)$, up to N, which can not be represented as*

$$n = p_1 + p_2^2 + p_3^3 + p_4^4. \tag{4}$$

Then, for any $\varepsilon > 0$, we have

$$E(N) \ll N^{\frac{61}{144}+\varepsilon}.$$

We will establish Theorem 1 by using a pruning process into the Hardy–Littlewood circle method. For the treatment on minor arcs, we will employ the argument developed by Wooley in [24] combined with the new estimates for exponential sum over primes developed by Zhao [25]. For the treatment on major arcs, we shall prune the major arcs further and deal with them respectively. The explicit details will be given in the related sections.

Notation. In this paper, let p, with or without subscripts, always denote a prime number; ε always denotes a sufficiently small positive constant, which may not be the same at different occurrences. The letter c always denotes a positive constant. As usual, we use $\chi \bmod q$ to denote a Dirichlet character modulo q, and $\chi^0 \bmod q$ the principal character. Moreover, we use $\varphi(n)$ and $d(n)$ to denote the Euler's function and Dirichlet's divisor function, respectively. $e(x) = e^{2\pi i x}$; $f(x) \ll g(x)$ means that $f(x) = O(g(x))$; $f(x) \asymp g(x)$ means that $f(x) \ll g(x) \ll f(x)$. N is a sufficiently large integer and $n \in (N/2, N]$, and hence $\log N \asymp \log n$.

2. Outline of the Proof of Theorem 1

Let N be a sufficiently large positive integer. By a splitting argument, it is sufficient to consider the even integers $n \in (N/2, N]$. For the application of the Hardy–Littlewood method, it is necessary to define the Farey dissection. For this purpose, we set the parameters as follows

$$A = 100^{100}, \quad Q_0 = \log^A N, \quad Q_1 = N^{\frac{1}{6}}, \quad Q_2 = N^{\frac{5}{6}}, \quad \mathfrak{I}_0 = \left[-\frac{1}{Q_2}, 1 - \frac{1}{Q_2} \right].$$

By Dirichlet's rational approximation lemma (for instance, see Lemma 12 on p.104 of [26], or Lemma 2.1 of [1]), each $\alpha \in (-1/Q_2, 1 - 1/Q_2]$ can be represented in the form

$$\alpha = \frac{a}{q} + \lambda, \qquad |\lambda| \leqslant \frac{1}{qQ_2},$$

for some integers a, q with $1 \leqslant a \leqslant q \leqslant Q_2$ and $(a, q) = 1$. Define

$$\mathfrak{M}(q, a) = \left[\frac{a}{q} - \frac{1}{qQ_2}, \frac{a}{q} + \frac{1}{qQ_2} \right], \qquad \mathfrak{M} = \bigcup_{1 \leqslant q \leqslant Q_1} \bigcup_{\substack{1 \leqslant a \leqslant q \\ (a,q)=1}} \mathfrak{M}(q, a),$$

$$\mathfrak{M}_0(q, a) = \left[\frac{a}{q} - \frac{Q_0^{100}}{qN}, \frac{a}{q} + \frac{Q_0^{100}}{qN} \right], \qquad \mathfrak{M}_0 = \bigcup_{1 \leqslant q \leqslant Q_0^{100}} \bigcup_{\substack{1 \leqslant a \leqslant q \\ (a,q)=1}} \mathfrak{M}_0(q, a),$$

$$\mathfrak{m}_1 = \mathfrak{I}_0 \setminus \mathfrak{M}, \qquad \mathfrak{m}_2 = \mathfrak{M} \setminus \mathfrak{M}_0.$$

Then we obtain the Farey dissection

$$\mathfrak{I}_0 = \mathfrak{M}_0 \cup \mathfrak{m}_1 \cup \mathfrak{m}_2. \tag{5}$$

For $k = 1, 2, 3, 4$, we define

$$f_k(\alpha) = \sum_{X_k < p \leqslant 2X_k} e(p^k \alpha),$$

where $X_k = (N/16)^{\frac{1}{k}}$. Let

$$\mathscr{R}(n) = \sum_{\substack{n = p_1 + p_2^2 + p_3^3 + p_4^4 \\ X_i < p_i \leqslant 2X_i \\ i=1,2,3,4}} 1.$$

From (5), one has

$$\mathscr{R}(n) = \int_0^1 \left(\prod_{k=1}^4 f_k(\alpha) \right) e(-n\alpha) d\alpha = \int_{-\frac{1}{Q_2}}^{1 - \frac{1}{Q_2}} \left(\prod_{k=1}^4 f_k(\alpha) \right) e(-n\alpha) d\alpha$$

$$= \left\{ \int_{\mathfrak{M}_0} + \int_{\mathfrak{m}_1} + \int_{\mathfrak{m}_2} \right\} \left(\prod_{k=1}^4 f_k(\alpha) \right) e(-n\alpha) d\alpha.$$

In order to prove Theroem 1, we need the two following propositions:

Proposition 1. *For $n \in (N/2, N]$, there holds*

$$\int_{\mathfrak{M}_0} \left(\prod_{k=1}^4 f_k(\alpha) \right) e(-n\alpha) d\alpha = \frac{\Gamma(2)\Gamma(\frac{3}{2})\Gamma(\frac{4}{3})\Gamma(\frac{5}{4})}{\Gamma(\frac{25}{12})} \mathfrak{S}(n) \frac{n^{\frac{13}{12}}}{\log^4 n} + O\left(\frac{n^{\frac{13}{12}}}{\log^5 n} \right), \tag{6}$$

where $\mathfrak{S}(n)$ is the singular series defined in (10), which is absolutely convergent and satisfies

$$(\log \log n)^{-c^*} \ll \mathfrak{S}(n) \ll d(n) \tag{7}$$

for any integer n satisfying $n \equiv 0 \,(\mathrm{mod}\, 2)$ and some fixed constant $c^* > 0$.

The proof of (6) in Proposition 1 follows from the well–know standard technique in the Hardy–Littlewood method. For more information, one can see pp. 90–99 of Hua [7], so we omit the details herein. For the properties (7) of singular series, we shall give the proof in Section 4.

Proposition 2. *Let $\mathscr{Z}(N)$ denote the number of integers $n \in (N/2, N]$ satisfying $n \equiv 0 \,(\mathrm{mod}\, 2)$ such that*

$$\sum_{j=1}^2 \left| \int_{\mathfrak{m}_j} \left(\prod_{k=1}^4 f_k(\alpha) \right) e(-n\alpha) d\alpha \right| \gg \frac{n^{\frac{13}{12}}}{\log^5 n}.$$

Then we have

$$\mathcal{Z}(N) \ll N^{\frac{61}{144}+\varepsilon}.$$

The proof of Proposition 2 will be given in Section 5. The remaining part of this section is devoted to establishing Theorem 1 by using Proposition 1 and Proposition 2.

Proof of Theorem 1. From Proposition 2, we deduce that, with at most $O\left(N^{\frac{61}{144}+\varepsilon}\right)$ exceptions, all even integers $n \in (N/2, N]$ satisfy

$$\sum_{j=1}^{2}\left|\int_{\mathfrak{m}_j}\left(\prod_{k=1}^{4} f_k(\alpha)\right)e(-n\alpha)\mathrm{d}\alpha\right| \ll \frac{n^{\frac{13}{12}}}{\log^5 n},$$

from which and Proposition 1, we conclude that, with at most $O\left(N^{\frac{61}{144}+\varepsilon}\right)$ exceptions, for all even integers $n \in (N/2, N]$, $\mathscr{R}(n)$ holds the asymptotic formula

$$\mathscr{R}(n) = \frac{\Gamma(2)\Gamma(\frac{3}{2})\Gamma(\frac{4}{3})\Gamma(\frac{5}{4})}{\Gamma(\frac{25}{12})}\mathfrak{S}(n)\frac{n^{\frac{13}{12}}}{\log^4 n} + O\left(\frac{n^{\frac{13}{12}}}{\log^5 n}\right).$$

In other words, all even integers $n \in (N/2, N]$ can be represented in the form $p_1 + p_2^2 + p_3^3 + p_4^4$ with at most $O\left(N^{\frac{61}{144}+\varepsilon}\right)$ exceptions, where p_1, p_2, p_3, p_4 are prime numbers. By a splitting argument, we get

$$E(N) \ll \sum_{0\leqslant \ell \ll \log N} \mathcal{Z}\left(\frac{N}{2^\ell}\right) \ll \sum_{0\leqslant \ell \ll \log N}\left(\frac{N}{2^\ell}\right)^{\frac{61}{144}+\varepsilon} \ll N^{\frac{61}{144}+\varepsilon}.$$

This completes the proof of Theorem 1.

3. Some Auxiliary Lemmas

In this section, we shall list some necessary lemmas which will be used in proving Proposition 2.

Lemma 1. *Suppose that α is a real number, and that $|\alpha - a/q| \leqslant q^{-2}$ with $(a, q) = 1$. Let $\beta = \alpha - a/q$. Then we have*

$$f_k(\alpha) \ll d^{\delta_k}(q)(\log x)^c\left(X_k^{1/2}\sqrt{q(1+N|\beta|)} + X_k^{4/5} + \frac{X_k}{\sqrt{q(1+N|\beta|)}}\right),$$

where $\delta_k = \frac{1}{2} + \frac{\log k}{\log 2}$ and c is a constant.

Proof. See Theorem 1.1 of Ren [27]. $\qquad\square$

Lemma 2. *Suppose that α is a real number, and that there exist $a \in \mathbb{Z}$ and $q \in \mathbb{N}$ with*

$$(a, q) = 1, \qquad 1 \leqslant q \leqslant X \qquad and \qquad |q\alpha - a| \leqslant X^{-1}.$$

If $P^{2\delta 2^{1-k}} \leqslant X \leqslant P^{k-2\delta 2^{1-k}}$, then one has

$$\sum_{P < p \leqslant 2P} e\left(p^k \alpha\right) \ll P^{1-\delta 2^{1-k}+\varepsilon} + \frac{P^{1+\varepsilon}}{q^{1/2}\left(1 + P^k|\alpha - a/q|\right)^{1/2}},$$

where $\delta = 1/3$ for $k \geqslant 4$.

Proof. See Lemma 2.4 of Zhao [25]. $\qquad\square$

Lemma 3. *Suppose that α is a real number, and that there are $a \in \mathbb{Z}$ and $q \in \mathbb{N}$ with*

$$(a,q) = 1, \qquad 1 \leqslant q \leqslant Q \qquad \text{and} \qquad |q\alpha - a| \leqslant Q^{-1}.$$

If $P^{\frac{1}{2}} \leqslant Q \leqslant P^{\frac{5}{2}}$, then one has

$$\sum_{P < p \leqslant 2P} e(p^3 \alpha) \ll P^{1 - \frac{1}{12} + \varepsilon} + \frac{q^{-\frac{1}{6}} P^{1+\varepsilon}}{\left(1 + P^3 |\alpha - a/q|\right)^{1/2}}.$$

Proof. See Lemma 8.5 of Zhao [25]. $\qquad\square$

Lemma 4. *For $\alpha \in \mathfrak{m}_1$, we have*

$$f_3(\alpha) \ll N^{\frac{11}{36} + \varepsilon} \qquad \text{and} \qquad f_4(\alpha) \ll N^{\frac{23}{96} + \varepsilon}.$$

Proof. For $\alpha \in \mathfrak{m}_1$, we have $Q_1 \leqslant q \leqslant Q_2$. By Lemma 3, we get

$$f_3(\alpha) \ll X_3^{\frac{11}{12} + \varepsilon} + X_3^{1+\varepsilon} Q_1^{-\frac{1}{6}} \ll N^{\frac{11}{36} + \varepsilon}.$$

From Lemma 2, we obtain

$$f_4(\alpha) \ll X_4^{\frac{23}{24} + \varepsilon} + X_4^{1+\varepsilon} Q_1^{-\frac{1}{2}} \ll N^{\frac{23}{96} + \varepsilon}.$$

This completes the proof of Lemma 4. $\qquad\square$

For $1 \leqslant a \leqslant q$ with $(a,q) = 1$, set

$$\mathcal{I}(q,a) = \left[\frac{a}{q} - \frac{1}{qQ_0}, \frac{a}{q} + \frac{1}{qQ_0} \right], \qquad \mathcal{I} = \bigcup_{\substack{1 \leqslant q \leqslant Q_0}} \bigcup_{\substack{a=-q \\ (a,q)=1}}^{2q} \mathcal{I}(q,a). \tag{8}$$

For $\alpha \in \mathfrak{m}_2$, by Lemma 1, we have

$$f_3(\alpha) \ll \frac{N^{\frac{1}{3}} \log^c N}{q^{\frac{1}{2} - \varepsilon} \left(1 + N|\lambda|\right)^{1/2}} + N^{\frac{4}{15} + \varepsilon} = V_3(\alpha) + N^{\frac{4}{15} + \varepsilon}, \tag{9}$$

say. Then we obtain the following Lemma.

Lemma 5. *We have*

$$\int_{\mathcal{I}} |V_3(\alpha)|^4 d\alpha = \sum_{\substack{1 \leqslant q \leqslant Q_0}} \sum_{\substack{a=-q \\ (a,q)=1}}^{2q} \int_{\mathcal{I}(q,a)} |V_3(\alpha)|^4 d\alpha \ll N^{\frac{1}{3}} \log^c N.$$

Proof. We have

$$
\sum_{\substack{1\leqslant q\leqslant Q_0}} \sum_{\substack{a=-q\\(a,q)=1}}^{2q} \int_{\mathcal{I}(q,a)} |V_3(\alpha)|^4 d\alpha
$$

$$
\ll \sum_{\substack{1\leqslant q\leqslant Q_0}} q^{-2+\varepsilon} \sum_{\substack{a=-q\\(a,q)=1}}^{2q} \int_{|\lambda|\leqslant\frac{1}{Q_0}} \frac{N^{\frac{4}{3}}\log^c N}{(1+N|\lambda|)^2} d\lambda
$$

$$
\ll \sum_{\substack{1\leqslant q\leqslant Q_0}} q^{-2+\varepsilon} \sum_{\substack{a=-q\\(a,q)=1}}^{2q} \left(\int_{|\lambda|\leqslant\frac{1}{N}} N^{\frac{4}{3}}\log^c N d\lambda + \int_{\frac{1}{N}\leqslant|\lambda|\leqslant\frac{1}{Q_0}} \frac{N^{\frac{4}{3}}\log^c N}{N^2\lambda^2} d\lambda \right)
$$

$$
\ll N^{\frac{1}{3}}\log^c N \sum_{\substack{1\leqslant q\leqslant Q_0}} q^{-2+\varepsilon}\varphi(q) \ll N^{\frac{1}{3}} Q_0^\varepsilon \log^c N \ll N^{\frac{1}{3}}\log^c N.
$$

This completes the proof of Lemma 5. $\qquad\square$

4. The Singular Series

In this section, we shall concentrate on investigating the properties of the singular series which appear in Proposition 1. First, we illustrate some notations. For $k \in \{1,2,3,4\}$ and a Dirichlet character χ mod q, we define

$$
C_k(\chi,a) = \sum_{h=1}^{q} \overline{\chi(h)} e\left(\frac{ah^k}{q}\right), \qquad C_k(q,a) = C_k(\chi^0,a),
$$

where χ^0 is the principal character modulo q. Let $\chi_1, \chi_2, \chi_3, \chi_4$ be Dirichlet characters modulo q. Set

$$
B(n,q,\chi_1,\chi_2,\chi_3,\chi_4) = \sum_{\substack{a=1\\(a,q)=1}}^{q} C_1(\chi_1,a)C_2(\chi_2,a)C_3(\chi_3,a)C_4(\chi_4,a)e\left(-\frac{an}{q}\right),
$$

$$
B(n,q) = B\left(n,q,\chi^0,\chi^0,\chi^0,\chi^0\right),
$$

and write

$$
A(n,q) = \frac{B(n,q)}{\varphi^4(q)}, \qquad \mathfrak{S}(n) = \sum_{q=1}^{\infty} A(n,q). \tag{10}
$$

Lemma 6. *For $(a,q)=1$ and any Dirichlet character χ mod q, there holds*

$$
|C_k(\chi,a)| \leqslant 2q^{1/2}d^{\beta_k}(q)
$$

with $\beta_k = (\log k)/\log 2$.

Proof. See the Problem 14 of Chapter VI of Vinogradov [28]. $\qquad\square$

Lemma 7. *Let p be a prime and $p^\alpha \| k$. For $(a,p)=1$, if $\ell \geqslant \gamma(p)$, we have $C_k(p^\ell,a)=0$, where*

$$
\gamma(p) = \begin{cases} \alpha+2, & \text{if } p \neq 2 \text{ or } p=2,\ \alpha=0; \\ \alpha+3, & \text{if } p=2,\ \alpha>0. \end{cases}
$$

Proof. See Lemma 8.3 of Hua [7]. $\qquad\square$

For $k \geqslant 1$, we define

$$
S_k(q,a) = \sum_{m=1}^{q} e\left(\frac{am^k}{q}\right).
$$

Lemma 8. *Suppose that* $(p, a) = 1$. *Then*

$$S_k(p, a) = \sum_{\chi \in \mathscr{A}_k} \overline{\chi(a)} \tau(\chi),$$

where $\mathscr{A}_k$ denotes the set of non–principal characters χ modulo p for which χ^k is principal, and $\tau(\chi)$ denotes the Gauss sum

$$\sum_{m=1}^{p} \chi(m) e\left(\frac{m}{p}\right).$$

Also, there hold $|\tau(\chi)| = p^{1/2}$ and $|\mathscr{A}_k| = (k, p-1) - 1$.

Proof. See Lemma 4.3 of Vaughan [1]. $\qquad\square$

Lemma 9. *For $(p, n) = 1$, we have*

$$\left| \sum_{a=1}^{p-1} \left(\prod_{k=1}^{4} \frac{S_k(p,a)}{p} \right) e\left(-\frac{an}{p}\right) \right| \leqslant 24 p^{-\frac{3}{2}}. \tag{11}$$

Proof. We denote by $\mathcal{S}$ the left-hand side of (11). It follows from Lemma 8 that

$$\mathcal{S} = \frac{1}{p^4} \sum_{a=1}^{p-1} \left(\prod_{k=1}^{4} \left(\sum_{\chi_k \in \mathscr{A}_k} \overline{\chi_k(a)} \tau(\chi_k) \right) \right) e\left(-\frac{an}{p}\right).$$

If $|\mathscr{A}_k| = 0$ for some $k \in \{1, 2, 3, 4\}$, then $\mathcal{S} = 0$. If this is not the case, then

$$\mathcal{S} = \frac{1}{p^4} \sum_{\chi_1 \in \mathscr{A}_1} \sum_{\chi_2 \in \mathscr{A}_2} \sum_{\chi_3 \in \mathscr{A}_3} \sum_{\chi_4 \in \mathscr{A}_4} \tau(\chi_1) \tau(\chi_2) \tau(\chi_3) \tau(\chi_4)$$

$$\times \sum_{a=1}^{p-1} \overline{\chi_1(a)\chi_2(a)\chi_3(a)\chi_4(a)} e\left(-\frac{an}{p}\right).$$

From Lemma 8, the quadruple outer sums have no more than $4! = 24$ terms. For each of these terms, there holds

$$|\tau(\chi_1)\tau(\chi_2)\tau(\chi_3)\tau(\chi_4)| = p^2.$$

Since in any one of these terms $\overline{\chi_1(a)\chi_2(a)\chi_3(a)\chi_4(a)}$ is a Dirichlet character $\chi \pmod p$, the inner sum is

$$\sum_{a=1}^{p-1} \chi(a) e\left(-\frac{an}{p}\right) = \overline{\chi(-n)} \sum_{a=1}^{p-1} \chi(-an) e\left(-\frac{an}{p}\right) = \overline{\chi(-n)} \tau(\chi).$$

By noting the fact that $\tau(\chi^0) = -1$ for principal character $\chi^0 \bmod p$, we derive that

$$\left| \overline{\chi(-n)} \tau(\chi) \right| \leqslant p^{\frac{1}{2}}.$$

From the above arguments, we deduce that

$$|\mathcal{S}| \leqslant \frac{1}{p^4} \cdot 24 \cdot p^2 \cdot p^{\frac{1}{2}} = 24 p^{-\frac{3}{2}},$$

which completes the proof of Lemma 9. $\qquad\square$

Lemma 10. *Let $\mathcal{L}(p, n)$ denote the number of solutions of the congruence*

$$x_1 + x_2^2 + x_3^3 + x_4^4 \equiv n \pmod p, \qquad 1 \leqslant x_1, x_2, x_3, x_4 \leqslant p - 1.$$

Then, for $n \equiv 0 \,(\mathrm{mod}\,2)$, we have $\mathcal{L}(p,n) > 0$.

Proof. We have

$$p \cdot \mathcal{L}(p,n) = \sum_{a=1}^{p} C_1(p,a)C_2(p,a)C_3(p,a)C_4(p,a)e\left(-\frac{an}{p}\right) = (p-1)^4 + E_p,$$

where

$$E_p = \sum_{a=1}^{p-1} C_1(p,a)C_2(p,a)C_3(p,a)C_4(p,a)e\left(-\frac{an}{p}\right).$$

By Lemma 8, we obtain

$$|E_p| \leqslant (p-1)(\sqrt{p}+1)(2\sqrt{p}+1)(3\sqrt{p}+1).$$

It is easy to check that $|E_p| < (p-1)^4$ for $p \geqslant 7$. Therefore, we obtain $\mathcal{L}(p,n) > 0$ for $p \geqslant 7$. For $p = 2,3,5$, we can check $\mathcal{L}(p,n) > 0$ one by one. This completes the proof of Lemma 10. $\quad\square$

Lemma 11. *$A(n,q)$ is multiplicative in q.*

Proof. From the definition of $A(n,q)$ in (10), it is sufficient to show that $B(n,q)$ is multiplicative in q. Suppose $q = q_1 q_2$ with $(q_1, q_2) = 1$. Then we obtain

$$\begin{aligned}
B(n, q_1 q_2) &= \sum_{\substack{a=1 \\ (a, q_1 q_2)=1}}^{q_1 q_2} \left(\prod_{k=1}^{4} C_k(q_1 q_2, a)\right) e\left(-\frac{an}{q_1 q_2}\right) \\
&= \sum_{\substack{a_1=1 \\ (a_1, q_1)=1}}^{q_1} \sum_{\substack{a_2=1 \\ (a_2, q_2)=1}}^{q_2} \left(\prod_{k=1}^{4} C_k(q_1 q_2, a_1 q_2 + a_2 q_1)\right) e\left(-\frac{a_1 n}{q_1}\right) e\left(-\frac{a_2 n}{q_2}\right).
\end{aligned} \tag{12}$$

For $(q_1, q_2) = 1$, there holds

$$\begin{aligned}
C_k(q_1 q_2, a_1 q_2 + a_2 q_1) &= \sum_{\substack{m=1 \\ (m, q_1 q_2)=1}}^{q_1 q_2} e\left(\frac{(a_1 q_2 + a_2 q_1)m^k}{q_1 q_2}\right) \\
&= \sum_{\substack{m_1=1 \\ (m_1, q_1)=1}}^{q_1} \sum_{\substack{m_2=1 \\ (m_2, q_2)=1}}^{q_2} e\left(\frac{(a_1 q_2 + a_2 q_1)(m_1 q_2 + m_2 q_1)^k}{q_1 q_2}\right) \\
&= \sum_{\substack{m_1=1 \\ (m_1, q_1)=1}}^{q_1} e\left(\frac{a_1(m_1 q_2)^k}{q_1}\right) \sum_{\substack{m_2=1 \\ (m_2, q_2)=1}}^{q_2} e\left(\frac{a_2(m_2 q_1)^k}{q_2}\right) \\
&= C_k(q_1, a_1) C_k(q_2, a_2).
\end{aligned} \tag{13}$$

Putting (13) into (12), we deduce that

$$\begin{aligned}
B(n, q_1 q_2) &= \sum_{\substack{a_1=1 \\ (a_1, q_1)=1}}^{q_1} \left(\prod_{k=1}^{4} C_k(q_1, a_1)\right) e\left(-\frac{a_1 n}{q_1}\right) \sum_{\substack{a_2=1 \\ (a_2, q_2)=1}}^{q_2} \left(\prod_{k=1}^{4} C_k(q_2, a_2)\right) e\left(-\frac{a_2 n}{q_2}\right) \\
&= B(n, q_1) B(n, q_2).
\end{aligned}$$

This completes the proof of Lemma 11. $\quad\square$

Lemma 12. *Let $A(n,q)$ be as defined in (10). Then*

(i)   we have

$$\sum_{q>Z} |A(n,q)| \ll Z^{-\frac{1}{2}+\varepsilon} d(n),$$

and thus the singular series $\mathfrak{S}(n)$ is absolutely convergent and satisfies $\mathfrak{S}(n) \ll d(n)$.

(ii)   there exists an absolute positive constant $c^* > 0$, such that, for $n \equiv 0 \,(\mathrm{mod}\,2)$,

$$\mathfrak{S}(n) \gg (\log \log n)^{-c^*}.$$

Proof. From Lemma 11, we know that $B(n,q)$ is multiplicative in q. Therefore, there holds

$$B(n,q) = \prod_{p^t \| q} B(n,p^t) = \prod_{p^t \| q} \sum_{\substack{a=1 \\ (a,p)=1}}^{p^t} \left(\prod_{k=1}^{4} C_k(p^t,a) \right) e\left(-\frac{an}{p^t} \right). \tag{14}$$

From (14) and Lemma 7, we deduce that $B(n,q) = \prod_{p\|q} B(n,p)$ or 0 according to q is square–free or not. Thus, one has

$$\sum_{q=1}^{\infty} A(n,q) = \sum_{\substack{q=1 \\ q \text{ square–free}}}^{\infty} A(n,q). \tag{15}$$

Write

$$\mathcal{R}(p,a) := \prod_{k=1}^{4} C_k(p,a) - \prod_{k=1}^{4} S_k(p,a).$$

Then

$$A(n,p) = \frac{1}{(p-1)^4} \sum_{a=1}^{p-1} \left(\prod_{k=1}^{4} S_k(p,a) \right) e\left(-\frac{an}{p} \right) + \frac{1}{(p-1)^4} \sum_{a=1}^{p-1} \mathcal{R}(p,a) e\left(-\frac{an}{p} \right). \tag{16}$$

Applying Lemma 6 and noticing that $S_k(p,a) = C_k(p,a) + 1$, we get $S_k(p,a) \ll p^{\frac{1}{2}}$, and thus $\mathcal{R}(p,a) \ll p^{\frac{3}{2}}$. Therefore, the second term in (16) is $\leqslant c_1 p^{-\frac{3}{2}}$. On the other hand, from Lemma 9, we can see that the first term in (16) is $\leqslant 2^4 \cdot 24 p^{-\frac{3}{2}} = 384 p^{-\frac{3}{2}}$. Let $c_2 = \max(c_1, 384)$. Then we have proved that, for $p \nmid n$, there holds

$$|A(n,p)| \leqslant c_2 p^{-\frac{3}{2}}. \tag{17}$$

Moreover, if we use Lemma 6 directly, it follows that

$$|B(n,p)| = \left| \sum_{a=1}^{p-1} \left(\prod_{k=1}^{4} C_k(p,a) \right) e\left(-\frac{an}{p} \right) \right| \leqslant \sum_{a=1}^{p-1} \prod_{k=1}^{4} |C_k(p,a)|$$
$$\leqslant (p-1) \cdot 2^4 \cdot p^2 \cdot 24 = 384 p^2 (p-1),$$

and therefore

$$|A(n,p)| = \frac{|B(n,p)|}{\varphi^4(p)} \leqslant \frac{384 p^2}{(p-1)^3} \leqslant \frac{2^3 \cdot 384 p^2}{p^3} = \frac{3072}{p}. \tag{18}$$

Let $c_3 = \max(c_2, 3072)$. Then, for square–free q, we have

$$|A(n,q)| = \left(\prod_{\substack{p|q \\ p\nmid n}} |A(n,p)| \right) \left(\prod_{\substack{p|q \\ p|n}} |A(n,p)| \right) \leqslant \left(\prod_{\substack{p|q \\ p\nmid n}} (c_3 p^{-\frac{3}{2}}) \right) \left(\prod_{\substack{p|q \\ p|n}} (c_3 p^{-1}) \right)$$
$$= c_3^{\omega(q)} \left(\prod_{p|q} p^{-\frac{3}{2}} \right) \left(\prod_{p|(n,q)} p^{\frac{1}{2}} \right) \ll q^{-\frac{3}{2}+\varepsilon} (n,q)^{\frac{1}{2}}.$$

Hence, by (15), we obtain

$$\sum_{q>Z} |A(n,q)| \ll \sum_{q>Z} q^{-\frac{3}{2}+\varepsilon}(n,q)^{\frac{1}{2}} = \sum_{d|n}\sum_{q>\frac{Z}{d}} (dq)^{-\frac{3}{2}+\varepsilon}d^{\frac{1}{2}} = \sum_{d|n} d^{-1+\varepsilon} \sum_{q>\frac{Z}{d}} q^{-\frac{3}{2}+\varepsilon}$$

$$\ll \sum_{d|n} d^{-1+\varepsilon}\left(\frac{Z}{d}\right)^{-\frac{1}{2}+\varepsilon} = Z^{-\frac{1}{2}+\varepsilon}\sum_{d|n} d^{-\frac{1}{2}+\varepsilon} \ll Z^{-\frac{1}{2}+\varepsilon}d(n).$$

This proves (i) of Lemma 12.

To prove (ii) of Lemma 12, by Lemma 11, we first note that

$$\mathfrak{S}(n) = \prod_p \left(1 + \sum_{t=1}^{\infty} A(n,p^t)\right) = \prod_p (1 + A(n,p))$$

$$= \left(\prod_{p \leqslant c_3} (1+A(n,p))\right)\left(\prod_{\substack{p>c_3 \\ p\nmid n}} (1+A(n,p))\right)\left(\prod_{\substack{p>c_3 \\ p\mid n}} (1+A(n,p))\right). \tag{19}$$

From (17), we have

$$\prod_{\substack{p>c_3 \\ p\nmid n}} (1+A(n,p)) \geqslant \prod_{p>c_3} \left(1 - \frac{c_3}{p^{3/2}}\right) \geqslant c_4 > 0. \tag{20}$$

By (18), we know that there are $c_5 > 0$ such that

$$\prod_{\substack{p>c_3 \\ p\mid n}} (1+A(n,p)) \geqslant \prod_{\substack{p>c_3 \\ p\mid n}} \left(1 - \frac{c_3}{p}\right) \geqslant \prod_{p\mid n} \left(1 - \frac{c_3}{p}\right) \gg (\log\log n)^{-c_5}. \tag{21}$$

On the other hand, it is easy to see that

$$1 + A(n,p) = \frac{p \cdot \mathcal{L}(p,n)}{\varphi^4(p)}.$$

By Lemma 10, we know that $\mathcal{L}(p,n) > 0$ for all p with $n \equiv 0 \,(\mathrm{mod}\,2)$, and thus $1 + A(n,p) > 0$. Therefore, there holds

$$\prod_{p\leqslant c_3} (1+A(n,p)) \geqslant c_6 > 0. \tag{22}$$

Combining the estimates (19)–(22), and taking $c^* = c_5 > 0$, we derive that

$$\mathfrak{S}(n) \gg (\log\log n)^{-c^*}.$$

This completes the proof Lemma 12. $\qquad\square$

5. Proof of Proposition 2

In this section, we shall give the proof of Proposition 2. We denote by $\mathcal{Z}_j(N)$ the set of integers n satisfying $n \in [N/2, N]$ and $n \equiv 0 \,(\mathrm{mod}\,2)$ for which the following estimate

$$\left|\int_{\mathfrak{m}_j} \left(\prod_{k=1}^{4} f_k(\alpha)\right) e(-n\alpha)\mathrm{d}\alpha\right| \gg \frac{n^{\frac{13}{12}}}{\log^5 n} \tag{23}$$

holds. For convenience, we use $\mathcal{Z}_j$ to denote the cardinality of $\mathcal{Z}_j(N)$ for abbreviation. Also, we define the complex number $\xi_j(n)$ by taking $\xi_j(n) = 0$ for $n \notin \mathcal{Z}_j(N)$, and when $n \in \mathcal{Z}_j(N)$ by means of the equation

$$\left| \int_{\mathfrak{m}_j} \left(\prod_{k=1}^{4} f_k(\alpha) \right) e(-n\alpha) d\alpha \right| = \xi_j(n) \int_{\mathfrak{m}_j} \left(\prod_{k=1}^{4} f_k(\alpha) \right) e(-n\alpha) d\alpha. \tag{24}$$

Plainly, one has $|\xi_j(n)| = 1$ whenever $\xi_j(n)$ is nonzero. Therefore, we obtain

$$\sum_{n \in \mathcal{Z}_j(N)} \xi_j(n) \int_{\mathfrak{m}_j} \left(\prod_{k=1}^{4} f_k(\alpha) \right) e(-n\alpha) d\alpha = \int_{\mathfrak{m}_j} \left(\prod_{k=1}^{4} f_k(\alpha) \right) \mathcal{K}_j(\alpha) d\alpha, \tag{25}$$

where the exponential sum $\mathcal{K}_j(\alpha)$ is defined by

$$\mathcal{K}_j(\alpha) = \sum_{n \in \mathcal{Z}_j(N)} \xi_j(n) e(-n\alpha).$$

For $j = 1, 2$, set

$$I_j = \int_{\mathfrak{m}_j} \left(\prod_{k=1}^{4} f_k(\alpha) \right) \mathcal{K}_j(\alpha) d\alpha.$$

By (23)–(25), we derive that

$$I_j \gg \sum_{n \in \mathcal{Z}_j(N)} \frac{n^{\frac{13}{12}}}{\log^5 n} \gg \frac{\mathcal{Z}_j N^{\frac{13}{12}}}{\log^5 N}, \qquad j = 1, 2. \tag{26}$$

By Lemma 2.1 of Wooley [24] with $k = 2$, we know that, for $j = 1, 2$, there holds

$$\int_0^1 |f_2(\alpha) \mathcal{K}_j(\alpha)|^2 d\alpha \ll N^\varepsilon (\mathcal{Z}_j N^{\frac{1}{2}} + \mathcal{Z}_j^2). \tag{27}$$

It follows from Cauchy's inequality, Lemma 4 and (27) that

$$I_1 \ll \left(\sup_{\alpha \in \mathfrak{m}_1} |f_3(\alpha)| \right) \left(\sup_{\alpha \in \mathfrak{m}_1} |f_4(\alpha)| \right) \left(\int_0^1 |f_2(\alpha) \mathcal{K}_1(\alpha)|^2 d\alpha \right)^{\frac{1}{2}} \left(\int_0^1 |f_1(\alpha)|^2 d\alpha \right)^{\frac{1}{2}}$$

$$\ll N^{\frac{11}{36}+\varepsilon} \cdot N^{\frac{23}{96}+\varepsilon} \cdot \left(N^\varepsilon (\mathcal{Z}_1 N^{\frac{1}{2}} + \mathcal{Z}_1^2) \right)^{\frac{1}{2}} \cdot N^{\frac{1}{2}}$$

$$\ll N^{\frac{301}{288}+\varepsilon} \left(\mathcal{Z}_1^{\frac{1}{2}} N^{\frac{1}{4}} + \mathcal{Z}_1 \right) \ll \mathcal{Z}_1^{\frac{1}{2}} N^{\frac{373}{288}+\varepsilon} + \mathcal{Z}_1 N^{\frac{301}{288}+\varepsilon}. \tag{28}$$

Combining (26) and (28), we get

$$\mathcal{Z}_1 N^{\frac{13}{12}} \log^{-5} N \ll I_1 \ll \mathcal{Z}_1^{\frac{1}{2}} N^{\frac{373}{288}+\varepsilon} + \mathcal{Z}_1 N^{\frac{301}{288}+\varepsilon},$$

which implies

$$\mathcal{Z}_1 \ll N^{\frac{61}{144}+\varepsilon}. \tag{29}$$

Next, we give the upper bound for $\mathcal{Z}_2$. By (9), we obtain

$$I_2 \ll \int_{\mathfrak{m}_2} |f_1(\alpha) f_2(\alpha) V_3(\alpha) f_4(\alpha) \mathcal{K}_2(\alpha)| d\alpha$$

$$+ N^{\frac{4}{15}+\varepsilon} \cdot \int_{\mathfrak{m}_2} |f_1(\alpha) f_2(\alpha) f_4(\alpha) \mathcal{K}_2(\alpha)| d\alpha$$

$$= I_{21} + I_{22}, \tag{30}$$

say. For $\alpha \in \mathfrak{m}_2$, we have either $Q_0^{100} < q < Q_1$ or $Q_0^{100} < N|q\alpha - a| < NQ_2^{-1} = Q_1$. Therefore, by Lemma 1, we get

$$\sup_{\alpha \in \mathfrak{m}_2} |f_4(\alpha)| \ll \frac{N^{\frac{1}{4}}}{\log^{40A} N}. \tag{31}$$

In view of the fact that $\mathfrak{m}_2 \subseteq \mathcal{I}$, where $\mathcal{I}$ is defined by (8), Hölder's inequality, the trivial estimate $\mathcal{K}_2(\alpha) \ll \mathcal{Z}_2$ and Theorem 4 of Hua (See [7], p. 19), we obtain

$$I_{21} \ll \mathcal{Z}_2 \sup_{\alpha \in \mathfrak{m}_2} |f_4(\alpha)| \times \left(\int_0^1 |f_1(\alpha)|^2 d\alpha \right)^{\frac{1}{2}} \left(\int_0^1 |f_2(\alpha)|^4 d\alpha \right)^{\frac{1}{4}} \left(\int_{\mathcal{I}} |V_3(\alpha)|^4 d\alpha \right)^{\frac{1}{4}}$$

$$\ll \mathcal{Z}_2 \cdot \frac{N^{\frac{1}{4}}}{\log^{40A} N} \cdot N^{\frac{1}{2}} \cdot (N \log^c N)^{\frac{1}{4}} \cdot (N^{\frac{1}{3}} \log^c N)^{\frac{1}{4}} \ll \frac{\mathcal{Z}_2 N^{\frac{13}{12}}}{\log^{30A} N}. \tag{32}$$

Moreover, it follows from (27), (31) and Cauchy's inequality that

$$I_{22} \ll N^{\frac{4}{15}+\varepsilon} \cdot \sup_{\alpha \in \mathfrak{m}_2} |f_4(\alpha)| \times \left(\int_0^1 |f_1(\alpha)|^2 d\alpha \right)^{\frac{1}{2}} \left(\int_0^1 |f_2(\alpha)\mathcal{K}_2(\alpha)|^2 d\alpha \right)^{\frac{1}{2}}$$

$$\ll N^{\frac{4}{15}+\varepsilon} \cdot \frac{N^{\frac{1}{4}}}{\log^{40A} N} \cdot N^{\frac{1}{2}} \cdot \left(N^\varepsilon \left(\mathcal{Z}_2 N^{\frac{1}{2}} + \mathcal{Z}_2^2 \right) \right)^{\frac{1}{2}}$$

$$\ll N^{\frac{61}{60}+\varepsilon} \left(\mathcal{Z}_2^{\frac{1}{2}} N^{\frac{1}{4}} + \mathcal{Z}_2 \right) \ll \mathcal{Z}_2^{\frac{1}{2}} N^{\frac{19}{15}+\varepsilon} + \mathcal{Z}_2 N^{\frac{61}{60}+\varepsilon}. \tag{33}$$

Combining (26), (30), (32) and (33), we deduce that

$$\frac{\mathcal{Z}_2 N^{\frac{13}{12}}}{\log^5 N} \ll I_2 = I_{21} + I_{22} \ll \frac{\mathcal{Z}_2 N^{\frac{13}{12}}}{\log^{30A} N} + \mathcal{Z}_2^{\frac{1}{2}} N^{\frac{19}{15}+\varepsilon} + \mathcal{Z}_2 N^{\frac{61}{60}+\varepsilon},$$

which implies

$$\mathcal{Z}_2 \ll N^{\frac{11}{30}+\varepsilon}. \tag{34}$$

From (29) and (34), we have

$$\mathcal{Z}(N) \ll \mathcal{Z}_1 + \mathcal{Z}_2 \ll N^{\frac{61}{144}+\varepsilon},$$

which completes the proof of Proposition 2.

Author Contributions: All authors contributed equally to this work. All authors have read and agreed to the published version of the manuscript.

Funding: This research work is supported by National Natural Science Foundation of China (Grant No. 11901566, 11971476), the Fundamental Research Funds for the Central Universities (Grant No. 2019QS02), and National Training Program of Innovation and Entrepreneurship for Undergraduates (Grant No. 201911413071, C201907622).

Acknowledgments: The authors would like to express the most sincere gratitude to the referee for his/her patience in refereeing this paper.

Conflicts of Interest: The authors declare no conflict of interest.

References

1. Vaughan, R.C. *The Hardy-Littlewood Method*; Cambridge University Press: Cambridge, UK, 1997.
2. Davenport, H. On Waring's problem for fourth powers. *Ann. Math.* **1939**, *40*, 731–747.
3. Hardy, G.H.; Ramanujan, S. Asymptotic formulae in combinatory analysis. *Proc. Lond. Math. Soc.* **1918**, *17*, 75–115.
4. Vinogradov, I.M. Representation of an odd number as a sum of three primes. *Dokl. Akad. Nauk SSSR* **1937**, *15*, 6–7.
5. Vinogradov, I.M. Some theorems concerning the theory of primes. *Mat. Sb.* **1937**, *44*, 179–195.

6.   Hua, L.K. *Additive Primzahltheorie*; B. G. Teubner Verlagsgesellschaft: Leipzig, Germany, 1959.
7.   Hua, L.K. *Additive Theory of Prime Numbers*; American Mathematical Society: Providence, RI, USA, 1965.
8.   Jagy, W.C.; Kaplansky, I. Sums of squares, cubes, and higher powers. *Exp. Math.* **1995**, *4*, 169–173.
9.   Brüdern, J. On Waring's problem for cubes and biquadrates. *J. Lond. Math. Soc.* **1988**, *37*, 25–42.
10. Brüdern, J.; Wooley, T.D. On Waring's problem: Three cubes and a sixth power. *Nagoya Math. J.* **2001**, *163*, 13–53.
11. Davenport, H.; Heilbronn, H. Note on a result in the additive theory of numbers. *Proc. Lond. Math. Soc.* **1937**, *2*, 142–151.
12. Davenport, H.; Heilbronn, H. On Waring's problem: Two cubes and one square. *Proc. Lond. Math. Soc.* **1937**, *2*, 73–104.
13. Roth, K.F. Proof that almost all positive integers are sums of a square, a positive cube and a fourth power. *J. Lond. Math. Soc.* **1949**, *24*, 4–13.
14. Vaughan, R.C. A ternary additive problem. *Proc. Lond. Math. Soc.* **1980**, *41*, 516–532.
15. Hooley, C. On a new approach to various problems of Waring's type. In *Recent Progress in Analytic Number Theory*; Academic Press: London, UK, 1981; Volume 1, pp. 127–191.
16. Davenport, H. On Waring's problem for cubes. *Acta Math.* **1939**, *71*, 123–143.
17. Lu, M.G. On Waring's problem for cubes and fifth power. *Sci. China Ser. A* **1993**, *36*, 641–662.
18. Kawada, K.; Wooley, T.D. Sums of fourth powers and related topics. *J. Reine Angew. Math.* **1999**, *512*, 173–223.
19. Vaughan, R.C. A new iterative method in Waring's problem. *Acta Math.* **1989**, *162*, 1–71.
20. Brüdern, J.; Kawada, K. Ternary problems in additive prime number theory. In *Analytic Number Theory*; Jia, C., Matsumoto, K., Eds.; Kluwer: Dordrecht, The Netherlands, 2002; pp. 39–91.
21. Zhao, L. On unequal powers of primes and powers of 2. *Acta Math. Hungar.* **2015**, *146*, 405–420.
22. Liu, Z.; Lü, G. On unlike powers of primes and powers of 2. *Acta Math. Hung.* **2011**, *132*, 125–139.
23. Lü, X.D. On unequal powers of primes and powers of 2. *Ramanujan J.* **2019**, *50*, 111–121.
24. Wooley, T.D. Slim exceptional sets and the asymptotic formula in Waring's problem. *Math. Proc. Camb. Philos. Soc.* **2003**, *134*, 193–206.
25. Zhao, L. On the Waring-Goldbach problem for fourth and sixth powers. *Proc. Lond. Math. Soc.* **2014**, *108*, 1593–1622.
26. Pan, C.D.; Pan, C.B. *Goldbach Conjecture*; Science Press: Beijing, China, 1981.
27. Ren, X.M. On exponential sums over primes and application in Waring–Goldbach problem. *Sci. China Ser. A* **2005**, *48*, 785–797.
28. Vinogradov, I.M. *Elements of Number Theory*; Dover Publications, New York, USA, 1954.

 *symmetry*

Article

Some Identities and Inequalities Involving Symmetry Sums of Legendre Polynomials

Tingting Wang * and Liang Qiao

College of Science, Northwest A&F University, Yangling 712100, China; 18309225762@163.com
* Correspondence: ttwang@nwsuaf.edu.cn

Received: 13 November 2019; Accepted: 12 December 2019; Published: 16 December 2019

Abstract: By using the analysis methods and the properties of Chebyshev polynomials of the first kind, this paper studies certain symmetry sums of the Legendre polynomials, and gives some new and interesting identities and inequalities for them, thus improving certain existing results.

Keywords: Legendre polynomials; Chebyshev polynomials of the first kind; power series; symmetry sums; polynomial identities; polynomial inequalities

---

1. Introduction

For any integer $n \geq 0$, the Legendre polynomials $\{P_n(x)\}$ are defined as follows:

$$P_n(x) = \frac{2n-1}{n} x P_{n-1}(x) - \frac{n-1}{n} P_{n-2}(x)$$

for all $n \geq 2$, with $P_0(x) = 1$ and $P_1(x) = x$, see [1,2] for more information.

The first few terms of $P_n(x)$ are $P_2(x) = \frac{1}{2}(3x^2 - 1)$, $P_3(x) = \frac{1}{2}(5x^3 - 3x)$, $P_4(x) = \frac{1}{8}(35x^4 - 30x^2 + 3)$, $P_5(x) = \frac{1}{8}(63x^5 - 70x^3 + 15x)$, $\cdots$.

In fact, the general term of $P_n(x)$ is given by the formula

$$P_n(x) = \frac{1}{2^n} \cdot \sum_{k=0}^{\left[\frac{n}{2}\right]} \frac{(-1)^k \cdot (2n-2k)!}{k! \cdot (n-k)! \cdot (n-2k)!} \cdot x^{n-2k},$$

where $[y]$ denotes the greatest integer less than or equal to y.

It is clear that $P_n(x)$ is an orthogonal polynomial (see [1,2]). That is,

$$\int_{-1}^{1} P_m(x) P_n(x) dx = \begin{cases} 0, & \text{if } m \neq n; \\ \dfrac{2}{2n+1}, & \text{if } m = n. \end{cases}$$

The generating function of $P_n(x)$ is

$$\frac{1}{\sqrt{1-2xt+t^2}} = \sum_{n=0}^{\infty} P_n(x) \cdot t^n, \ |x| \leq 1, \ |t| < 1. \tag{1}$$

These polynomials play a vital role in the study of function orthogonality and approximation theory, as a result, some scholars have dedicated themselves to studying their various natures and obtained a series of meaningful research results. The studies that are concerned with this content can be found in [1–20]. Recently, Shen Shimeng and Chen Li [3] give certain symmetry sums of $P_n(x)$, and proved the following result:

For any positive integer k and integer $n \geq 0$, one has the identity

$$(2k-1)!! \sum_{a_1+a_2+\cdots+a_{2k+1}=n} P_{a_1}(x) P_{a_2}(x) \cdots P_{a_{2k+1}}(x)$$

$$= \sum_{j=1}^{k} C(k,j) \sum_{i=0}^{n} \frac{(n+k+1-i-j)!}{(n-i)!} \cdot \frac{\binom{i+j+k-2}{i}}{x^{k-1+i+j}} \cdot P_{n+k+1-i-j}(x),$$

where $(2k-1)!! = (2k-1)\cdot(2k-3)\cdots 3\cdot 1$, and $C(k,i)$ is a recurrence sequence defined by $C(k,1) = 1$, $C(k+1,k+1) = (2k-1)!!$ and $C(k+1,i+1) = C(k,i+1) + (k-1+i)\cdot C(k,i)$ for all $1 \leq i \leq k-1$.

The calculation formula for the sum of Legendre polynomials given above is virtually a linear combination of some $P_n(x)$, and the coefficients $C(k,i)$ are very regular. However, the result is in the form of a recursive formula, in other words, especially when k is relatively large, the formula is not actually easy to use for calculating specific values.

In an early paper, Zhou Yalan and Wang Xia [4] obtained some special cases with $k = 3$ and $k = 5$. It is even harder to calculate their exact values for the general positive integer k, especially if k is large enough.

Naturally, we want to ask a question: Is there a more concise and specific formula for the calculation of the above problems? This is the starting point of this paper. We used the different methods to come up with additional simpler identities. It is equal to saying that we have used the analysis method and the properties of the first kind of Chebyshev polynomials, thereby establishing the symmetry of the Legendre polynomial and symmetry relationship with the first kind of Chebyshev polynomial, and proved the following three results:

Theorem 1. *For any integers $k \geq 1$ and $n \geq 0$, we have the identity*

$$\sum_{a_1+a_2+\cdots+a_k=n} P_{a_1}(x) \cdot P_{a_2}(x) \cdots P_{a_k}(x) = \sum_{i=0}^{n} \frac{<\frac{k}{2}>_i}{i!} \cdot \frac{<\frac{k}{2}>_{n-i}}{(n-i)!} \cdot T_{n-2i}(x),$$

where $< x >_0 = 1$, $< x >_k = x(x+1)(x+2)\cdots(x+k-1)$ for all integers $k \geq 1$, and $T_n(x) = T_{-n}(x) = \frac{1}{2}\left(\left(x+\sqrt{x^2-1}\right)^n + \left(x+\sqrt{x^2-1}\right)^n\right)$ denotes Chebyshev polynomials of the first kind.

Theorem 2. *Let $q > 1$ is an integer, χ is any primitive character $\bmod q$. Then for any integers $k \geq 1$ and $n \geq 0$, we have the inequality*

$$\left| \sum_{a_1+a_2+\cdots+a_k=n} \sum_{a=1}^{q} \chi(a) P_{a_1}\left(\cos\frac{2\pi a}{q}\right) \cdot P_{a_2}\left(\cos\frac{2\pi a}{q}\right) \cdots P_{a_k}\left(\cos\frac{2\pi a}{q}\right) \right|$$

$$\leq \sqrt{q} \cdot \binom{n+k-1}{k-1}.$$

Theorem 3. *For any integer $n \geq 0$ with $2 \nmid n$, we have the identity*

$$\int_{-\frac{\pi}{2}}^{\frac{\pi}{2}} \left(\sum_{a_1+a_2+\cdots+a_k=n} P_{a_1}(\sin\theta) \cdot P_{a_2}(\sin\theta) \cdots P_{a_k}(\sin\theta) \right)^2 d\theta$$

$$= 2\pi \cdot \sum_{i=0}^{\left[\frac{n}{2}\right]} \left(\frac{<\frac{k}{2}>_i}{i!} \cdot \frac{<\frac{k}{2}>_{n-i}}{(n-i)!} \right)^2;$$

If $n = 2m$, then we have

$$\int_{-\frac{\pi}{2}}^{\frac{\pi}{2}} \left(\sum_{a_1+a_2+\cdots+a_k=n} P_{a_1}\left(\sin\theta\right) \cdot P_{a_2}\left(\sin\theta\right) \cdots P_{a_k}\left(\sin\theta\right) \right)^2 d\theta$$

$$= 2\pi \cdot \sum_{i=0}^{m} \left(\frac{<\frac{k}{2}>_i}{i!} \cdot \frac{<\frac{k}{2}>_{2m-i}}{(2m-i)!} \right)^2 - \pi \cdot \left(\frac{<\frac{k}{2}>_m}{m!} \right)^4.$$

Essentially, the main result of this paper is Theorem 1, which not only reveals the profound properties of Legendre polynomials and Chebyshev polynomials, but also greatly simplifies the calculation of the symmetry sum of Legendre polynomials in practice. We can replace the calculation of the symmetric sum of the Legendre polynomial with the first single Chebyshev polynomial calculation, which can greatly simplify the calculation of the symmetric sum.

Theorem 2 gives an upper bound estimate of the character sum of Legendre polynomials. Theorem 3 reveals the orthogonality of the symmetry sum of Legendre polynomials, which is a generalization of the orthogonality of functions. Of course, Theorems 2 and 3 can also be seen as the direct application of Theorem 1 in analytical number theory and the orthogonality of functions. This is of great significance in analytic number theory, and it has also made new contributions to the study of Gaussian sums.

In fact if we taking $k = 1$, and note that the identity $\frac{<\frac{1}{2}>_h}{h!} = \frac{1}{4^h} \cdot \binom{2h}{h}$, then from our theorems we may immediately deduce the following three corollaries.

Corollary 1. *For any integer $n \geq 0$, we have the identity*

$$P_n(x) = \frac{1}{4^n} \sum_{i=0}^{n} \binom{2i}{i}\binom{2n-2i}{n-i} \cdot T_{n-2i}(x),$$

where $T_n(x)$ denotes Chebyshev polynomials of the first kind.

Corollary 2. *Let $q > 1$ is an integer, χ is any primitive character $\bmod q$. Then for any integer $n \geq 0$, we have the inequality*

$$\left| \sum_{a=1}^{q} \chi(a) P_n\left(\cos\frac{2\pi a}{q} \right) \right| \leq \sqrt{q}.$$

Corollary 3. *For any integer $n \geq 0$ with $2 \nmid n$, we have the identity*

$$\int_{-\frac{\pi}{2}}^{\frac{\pi}{2}} P_n^2\left(\sin\theta\right) d\theta = \frac{2\pi}{4^{2n}} \sum_{i=0}^{\left[\frac{n}{2}\right]} \binom{2i}{i}^2 \binom{2n-2i}{n-i}^2;$$

If $n = 2m$, then we have the identity

$$\int_{-\frac{\pi}{2}}^{\frac{\pi}{2}} P_n^2\left(\sin\theta\right) d\theta = \frac{2\pi}{4^{2n}} \sum_{i=0}^{\left[\frac{n}{2}\right]} \binom{2i}{i}^2 \binom{2n-2i}{n-i}^2 - \frac{\pi}{4^{2n}} \cdot \binom{2m}{m}^4.$$

2. Proofs of the Theorems

In this section, we will directly prove the main results in this paper by by means of the properties of characteristic roots.

Proof of Theorem 1. First we prove Theorem 1. Let $\alpha = x + \sqrt{x^2 - 1}$ and $\beta = x - \sqrt{x^2 - 1}$ be two characteristic roots of the characteristic equation $\lambda^2 - 2x\lambda + 1 = 0$. Then from the definition and properties of Chebyshev polynomials $T_n(x)$ of the first kind, we have

$$T_n(x) = T_{-n}(x) = \frac{1}{2}\left(\alpha^n + \beta^n\right), \quad n \geq 0.$$

For any positive integer k, combining properties of power series and Formula (1) we have the identity

$$\left(\frac{1}{\sqrt{1-2xt+t^2}}\right)^k = \frac{1}{(1-2xt+t^2)^{\frac{k}{2}}} = \frac{1}{(1-\alpha t)^{\frac{k}{2}}(1-\beta t)^{\frac{k}{2}}}$$
$$= \sum_{n=0}^{\infty}\left(\sum_{a_1+a_2+\cdots+a_k=n} P_{a_1}(x)\cdot P_{a_2}(x)\cdot P_{a_3}(x)\cdots P_{a_k}(x)\right)\cdot t^n. \tag{2}$$

At the same time, we focus on the power series

$$\frac{1}{(1-x)^{\frac{k}{2}}} = \sum_{n=0}^{\infty}\frac{<\frac{k}{2}>_n}{n!}\cdot x^n, \quad |x|<1, \tag{3}$$

where $<x>_0=1$, $<x>_h= x(x+1)(x+2)\cdots(x+h-1)$ for all integers $h\geq 1$.

So for any positive integer k, note that $\alpha\cdot\beta=1$, from (3) and the symmetry properties of α and β we have

$$\left(\frac{1}{\sqrt{1-2xt+t^2}}\right)^k = \frac{1}{(1-\alpha t)^{\frac{k}{2}}(1-\beta t)^{\frac{k}{2}}}$$
$$= \left(\sum_{n=0}^{\infty}\frac{<\frac{k}{2}>_n}{n!}\cdot \alpha^n\cdot t^n\right)\left(\sum_{n=0}^{\infty}\frac{<\frac{k}{2}>_n}{n!}\cdot \beta^n\cdot t^n\right)$$
$$= \sum_{n=0}^{\infty}\left(\sum_{i=0}^{n}\frac{<\frac{k}{2}>_i}{i!}\cdot\frac{<\frac{k}{2}>_{n-i}}{(n-i)!}\cdot \alpha^i\cdot\beta^{n-i}\right)\cdot t^n$$
$$= \sum_{n=0}^{\infty}\left(\sum_{i=0}^{n}\frac{<\frac{k}{2}>_i}{i!}\cdot\frac{<\frac{k}{2}>_{n-i}}{(n-i)!}\cdot \beta^{n-2i}\right)\cdot t^n$$
$$= \sum_{n=0}^{\infty}\left(\sum_{i=0}^{n}\frac{<\frac{k}{2}>_i}{i!}\cdot\frac{<\frac{k}{2}>_{n-i}}{(n-i)!}\cdot \alpha^{n-2i}\right)\cdot t^n$$
$$= \sum_{n=0}^{\infty}\left(\sum_{i=0}^{n}\frac{<\frac{k}{2}>_i}{i!}\cdot\frac{<\frac{k}{2}>_{n-i}}{(n-i)!}\cdot\frac{1}{2}\left(\alpha^{n-2i}+\beta^{n-2i}\right)\right)\cdot t^n$$
$$= \sum_{n=0}^{\infty}\left(\sum_{i=0}^{n}\frac{<\frac{k}{2}>_i}{i!}\cdot\frac{<\frac{k}{2}>_{n-i}}{(n-i)!}\cdot T_{n-2i}(x)\right)\cdot t^n. \tag{4}$$

Combining (2) and (4), and then by comparing the coefficients on both sides of the power series, we can find

$$\sum_{a_1+a_2+\cdots+a_k=n} P_{a_1}(x)\cdot P_{a_2}(x)\cdots P_{a_k}(x) = \sum_{i=0}^{n}\frac{<\frac{k}{2}>_i}{i!}\cdot\frac{<\frac{k}{2}>_{n-i}}{(n-i)!}\cdot T_{n-2i}(x).$$

This proves Theorem 1. $\quad\square$

Proof of Theorem 2. The proof of Theorem 2 is next. Let $q>1$ be any integer, χ denotes any primitive character mod q. Then from Theorem 1 with $x = \cos\left(\frac{2\pi a}{q}\right)$ and the identity $T_n(\cos\theta) = \cos(n\theta)$, we have

$$\sum_{a_1+a_2+\cdots+a_k=n} \sum_{a=1}^{q} \chi(a) P_{a_1}\left(\cos\frac{2\pi a}{q}\right) \cdot P_{a_2}\left(\cos\frac{2\pi a}{q}\right) \cdots P_{a_k}\left(\cos\frac{2\pi a}{q}\right)$$

$$= \sum_{i=0}^{n} \frac{<\frac{k}{2}>_i}{i!} \cdot \frac{<\frac{k}{2}>_{n-i}}{(n-i)!} \cdot \sum_{a=1}^{q} \chi(a) \cdot \cos\left(\frac{2\pi a(n-2i)}{q}\right)$$

$$= \frac{1}{2} \sum_{i=0}^{n} \frac{<\frac{k}{2}>_i}{i!} \cdot \frac{<\frac{k}{2}>_{n-i}}{(n-i)!} \cdot \sum_{a=1}^{q} \chi(a) \left(e\left(\frac{a(n-2i)}{q}\right) + e\left(\frac{-a(n-2i)}{q}\right)\right)$$

$$= \frac{\tau(\chi)}{2} \sum_{i=0}^{n} \frac{<\frac{k}{2}>_i}{i!} \cdot \frac{<\frac{k}{2}>_{n-i}}{(n-i)!} \cdot \left(\overline{\chi}(n-2i) + \overline{\chi}(-(n-2i))\right), \tag{5}$$

where $e(y) = e^{2\pi i y}$, and $\sum_{a=1}^{q} \chi(a) e\left(\frac{na}{q}\right) = \overline{\chi}(n)\tau(\chi)$.

Note that for any primitive character $\chi \bmod q$, from the properties of Gauss sums, we have $|\tau(\chi)| = \sqrt{q}$, and for any positive integer $k \geq 1$, we have

$$\frac{1}{(1-x)^k} = \sum_{n=0}^{\infty} \binom{n+k-1}{k-1} \cdot x^n = \frac{1}{(1-x)^{\frac{k}{2}}} \cdot \frac{1}{(1-x)^{\frac{k}{2}}}$$

$$= \sum_{n=0}^{\infty} \left(\sum_{i=0}^{n} \frac{<\frac{k}{2}>_i}{i!} \cdot \frac{<\frac{k}{2}>_{n-i}}{(n-i)!}\right) \cdot x^n$$

or

$$\binom{n+k-1}{k-1} = \sum_{i=0}^{n} \frac{<\frac{k}{2}>_i}{i!} \cdot \frac{<\frac{k}{2}>_{n-i}}{(n-i)!}. \tag{6}$$

Combining (5) and (6), there will be an estimation formula immediately deduced

$$\left|\sum_{a_1+a_2+\cdots+a_k=n} \sum_{a=1}^{q} \chi(a) P_{a_1}\left(\cos\frac{2\pi a}{q}\right) \cdot P_{a_2}\left(\cos\frac{2\pi a}{q}\right) \cdots P_{a_k}\left(\cos\frac{2\pi a}{q}\right)\right|$$

$$= \frac{\sqrt{q}}{2} \cdot \left|\sum_{i=0}^{n} \frac{<\frac{k}{2}>_i}{i!} \cdot \frac{<\frac{k}{2}>_{n-i}}{(n-i)!} \cdot \left[\overline{\chi}(n-2i) + \overline{\chi}(-(n-2i))\right]\right|$$

$$\leq \sqrt{q} \cdot \sum_{i=0}^{n} \frac{<\frac{k}{2}>_i}{i!} \cdot \frac{<\frac{k}{2}>_{n-i}}{(n-i)!} = \sqrt{q} \cdot \binom{n+k-1}{k-1}.$$

Theorem 2 is proven completely. $\square$

Proof of Theorem 3. We prove Theorem 3 below. From the orthogonality of Chebyshev polynomials of the first kind we know that

$$\int_{-1}^{1} \frac{T_m(x) T_n(x)}{\sqrt{1-x^2}} dx = \begin{cases} 0, & \text{if } m \neq n; \\ \dfrac{\pi}{2}, & \text{if } m = n > 0, \\ \pi, & \text{if } m = n = 0. \end{cases} \tag{7}$$

If integer $n \geq 1$ with $2 \nmid n$, then for any integer $0 \leq i \leq n$, we have $n - 2i \neq 0$, note that $T_n(x) = T_{-n}(x)$, so from (7) and Theorem 1 we have

$$\int_{-1}^{1} \frac{1}{\sqrt{1-x^2}} \left(\sum_{a_1+a_2+\cdots+a_k=n} P_{a_1}(x) \cdot P_{a_2}(x) \cdots P_{a_k}(x) \right)^2 dx$$

$$= \int_{-1}^{1} \frac{1}{\sqrt{1-x^2}} \left(\sum_{i=0}^{n} \frac{<\frac{k}{2}>_i}{i!} \cdot \frac{<\frac{k}{2}>_{n-i}}{(n-i)!} \cdot T_{n-2i}(x) \right)^2 dx$$

$$= 4 \int_{-1}^{1} \frac{1}{\sqrt{1-x^2}} \left(\sum_{i=0}^{\left[\frac{n}{2}\right]} \frac{<\frac{k}{2}>_i}{i!} \cdot \frac{<\frac{k}{2}>_{n-i}}{(n-i)!} \cdot T_{n-2i}(x) \right)^2 dx$$

$$= 2\pi \cdot \sum_{i=0}^{\left[\frac{n}{2}\right]} \left(\frac{<\frac{k}{2}>_i}{i!} \cdot \frac{<\frac{k}{2}>_{n-i}}{(n-i)!} \right)^2. \tag{8}$$

For $n = 2m$, if $n - 2i = 0$, then $i = m$. So from (7), Theorem 1 and the methods of proving (8) we have

$$\int_{-1}^{1} \frac{1}{\sqrt{1-x^2}} \left(\sum_{a_1+a_2+\cdots+a_k=n} P_{a_1}(x) \cdot P_{a_2}(x) \cdots P_{a_k}(x) \right)^2 dx$$

$$= \int_{-1}^{1} \frac{1}{\sqrt{1-x^2}} \left(\left(\frac{<\frac{k}{2}>_m}{m!} \right)^2 + 2 \sum_{i=0}^{m-1} \frac{<\frac{k}{2}>_i}{i!} \cdot \frac{<\frac{k}{2}>_{2m-i}}{(2m-i)!} \cdot T_{2m-2i}(x) \right)^2 dx$$

$$= \pi \cdot \left(\frac{<\frac{k}{2}>_m}{m!} \right)^4 + 2\pi \cdot \sum_{i=0}^{m-1} \left(\frac{<\frac{k}{2}>_i}{i!} \cdot \frac{<\frac{k}{2}>_{n-i}}{(n-i)!} \right)^2$$

$$= 2\pi \cdot \sum_{i=0}^{m} \left(\frac{<\frac{k}{2}>_i}{i!} \cdot \frac{<\frac{k}{2}>_{2m-i}}{(2m-i)!} \right)^2 - \pi \cdot \left(\frac{<\frac{k}{2}>_m}{m!} \right)^4. \tag{9}$$

Let $x = \sin\theta$, then we have

$$\int_{-1}^{1} \frac{1}{\sqrt{1-x^2}} \left(\sum_{a_1+a_2+\cdots+a_k=n} P_{a_1}(x) \cdot P_{a_2}(x) \cdots P_{a_k}(x) \right)^2 dx$$

$$= \int_{-\frac{\pi}{2}}^{\frac{\pi}{2}} \left(\sum_{a_1+a_2+\cdots+a_k=n} P_{a_1}(\sin\theta) \cdot P_{a_2}(\sin\theta) \cdots P_{a_k}(\sin\theta) \right)^2 d\theta. \tag{10}$$

Now Theorem 3 follows from (8), (9), and (10). □

3. Conclusions

Three theorems and three inferences are the main results in the paper. Theorem 1 gives proof of the symmetry of Legendre polynomials and the symmetry relationship with Chebyshev polynomials of the first kind. This conclusion also improves the early results in [4], and also gives us a different representation for the result in [3]. Theorem 2 obtained an inequality involving Dirichlet characters and Legendre polynomials; this is actually a new contribution to the study of Legendre polynomials and character sums mod q. Theorem 3 established an integral identity involving the symmetry sums of the Legendre polynomials. The three corollaries are some special cases of our three theorems for $k = 1$, and can not only enrich the research content of the Legendre polynomials, but also promote its research development.

Author Contributions: All authors have equally contributed to this work. All authors read and approved the final manuscript.

Funding: This work is supported by the Y. S. T. N. S. P (2019KJXX-076) and N. S. B. R. P in Shaanxi Province (2019JM-207).

Acknowledgments: The authors would like to thank the editor and referee for their very helpful and detailed comments, which have significantly improved the presentation of this paper.

Conflicts of Interest: The authors declare that there are no conflicts of interest regarding the publication of this paper.

References

1. Borwein, P.; Erdèlyi, T. *Polynomials and Polynomial Inequalities*; Springer: New York, NY, USA, 1995.
2. Jackson, D. *Fourier Series and Orthogonal Polynomials*; Dover Publications: Mineola, NY, USA, 2004.
3. Shen, S.M.; Chen, L. Some types of identities involving the Legendre polynomials. *Mathematics* **2019**, *7*, 114. [CrossRef]
4. Zhou, Y.L.; Wang, X. The relationship of Legendre polynomials and Chebyshev polynomials. *Pure Appl. Math.* **1999**, *15*, 75–81.
5. Zhang, Y.X.; Chen, Z.Y. A new identity involving the Chebyshev polynomials. *Mathematics* **2018**, *6*, 244. [CrossRef]
6. Kim, T.; Kim, D.S.; Dolgy, D.V. Sums of finite products of Legendre and Laguerre polynomials. *Adv. Differ. Equ.* **2018**, *2018*, 277. [CrossRef]
7. Wang, S.Y. Some new identities of Chebyshev polynomials and their applications. *Adv. Differ. Equ.* **2015**, *2015*, 355.
8. He, Y. Some results for sums of products of Chebyshev and Legendre polynomials. *Adv. Differ. Equ.* **2019**, *2019*, 357. [CrossRef]
9. Chen, L.; Zhang, W.P. Chebyshev polynomials and their some interesting applications. *Adv. Differ. Equ.* **2017**, *2017*, 303.
10. Li, X.X. Some identities involving Chebyshev polynomials. *Math. Probl. Eng.* **2015**, *2015*, 950695. [CrossRef]
11. Wang, T.T.; Zhang, H. Some identities involving the derivative of the first kind Chebyshev polynomials. *Math. Probl. Eng.* **2015**, *2015*, 146313. [CrossRef]
12. Zhang, W.P.; Wang, T.T. Two identities involving the integral of the first kind Chebyshev polynomials. *Bull. Math. Soc. Sci. Math. Roum.* **2017**, *108*, 91–98.
13. Kim, T.; Dolgy, D.V.; Kim, D.S. Representing sums of finite products of Chebyshev polynomials of the second kind and Fibonacci polynomials in terms of Chebyshev polynomials. *Adv. Stud. Contemp. Math.* **2018**, *28*, 321–336.
14. Cesarano, C. Identities and generating functions on Chebyshev polynomials. *Georgian Math. J.* **2012**, *19*, 427–440. [CrossRef]
15. Cesarano, C. Generalized Chebyshev polynomials. *Hacet. J. Math. Stat.* **2014**, *43*, 731–740.
16. Dattoli, G.; Srivastava, H.M.; Cesarano, C. The Laguerre and Legendre polynomials from an operational point of view. *Appl. Math. Comput.* **2001**, *124*, 117–127. [CrossRef]
17. Islam, S.; Hossain, B. Numerical solutions of eighth order BVP by the Galerkin residual technique with Bernstein and Legendre polynomials. *Appl. Math. Comput.* **2015**, *261*, 48–59. [CrossRef]
18. Khalil, H.; Rahmat, A.K. A new method based on Legendre polynomials for solutions of the fractional two-dimensional heat conduction equation. *Comput. Math. Appl.* **2014**, *67*, 1938–1953. [CrossRef]
19. Wan, J.; Zudilin, W. Generating functions of Legendre polynomials: A tribute to Fred Brafman. *J. Approx. Theory* **2013**, *170*, 198–213. [CrossRef]
20. Nemati, S.; Lima, P. M.; Ordokhani, Y. Numerical solution of a class of two-dimensional nonlinear Volterra integral equations using Legendre polynomials. *J. Comput. Appl. Math.* **2013**, *242*, 53–69. [CrossRef]

 *symmetry*

Article

A New Identity Involving Balancing Polynomials and Balancing Numbers

Yuanyuan Meng

School of Mathematics, Northwest University, Xi'an 710127, Shaanxi, China; yymeng@stumail.nwu.edu.cn

Received: 13 August 2019 ; Accepted: 5 September 2019 ; Published: 7 September 2019

Abstract: In this paper, a second-order nonlinear recursive sequence $M(h, i)$ is studied. By using this sequence, the properties of the power series, and the combinatorial methods, some interesting symmetry identities of the structural properties of balancing numbers and balancing polynomials are deduced.

Keywords: balancing numbers; balancing polynomials; combinatorial methods; symmetry sums

1. Introduction

For any positive integer $n \geq 2$, we denote the balancing number by B_n and the balancer corresponding to it by $r(n)$ if

$$1 + 2 + \cdots + (B_n - 1) = (B_n + 1) + (B_n + 2) + \cdots + (B_n + r(n))$$

holds for some positive integer $r(n)$ and B_n. It is clear that $r(n) = \frac{B_n - B_{n-1} - 1}{2}$, for example, $r(2) = 2$, $r(3) = 14, r(4) = 84, r(5) = 492\ldots$

It is found that the balancing numbers satisfy the second order linear recursive sequence $B_{n+1} = 6B_n - B_{n-1}$ $(n \geq 1)$, providing $B_0 = 0$ and $B_1 = 1$ [1].

The balancing polynomials $B_n(x)$ are defined by $B_0(x) = 1$, $B_1(x) = 6x$, $B_2(x) = 36x^2 - 1$, $B_3(x) = 216x^3 - 12x, B_4(x) = 1296x^4 - 108x^2 + 1$, and the second-order linear difference equation:

$$B_{n+1}(x) = 6xB_n(x) - B_{n-1}(x), n \geq 1,$$

where x is any real number. While $n \geq 1$, we get $B_{n+1} = 6B_n - B_{n-1}$ with $B_n(1) = B_{n+1}$. Such balancing numbers have been widely studied in recent years. G. K. Panda and T. Komatsu [2] studied the reciprocal sums of the balancing numbers and proved the following inequation holds for any positive integer n:

$$\frac{1}{B_n - B_{n-1}} < \sum_{k=n}^{\infty} \frac{1}{B_k} < \frac{1}{B_n - B_{n-1} - 1}.$$

G. K. Panda [3] studied some fascinating properties of balancing numbers and gave the following result for any natural numbers $m > n$:

$$(B_m + B_n)(B_m - B_n) = B_{m+n} \cdot B_{m-n}.$$

Other achievements related to balancing numbers can be found in [4–7].

It is found that the balancing polynomials $B_n(x)$ can be generally expressed as

$$B_n(x) = \frac{1}{2\sqrt{9x^2 - 1}} \left[\left(3x + \sqrt{9x^2 - 1} \right)^{n+1} - \left(3x - \sqrt{9x^2 - 1} \right)^{n+1} \right],$$

and the generating function of the balancing polynomials $B_n(x)$ is given by

$$\frac{1}{1 - 6xt + t^2} = \sum_{n=0}^{\infty} B_n(x) \cdot t^n. \tag{1}$$

Recently, our attention was drawn to the sums of polynomials calculating problem [8–11], which is important in mathematical application. We are going to study the computational problem of the symmetry summation:

$$\sum_{a_1 + a_2 + \cdots + a_{h+1} = n} B_{a_1}(x) B_{a_2}(x) \cdots B_{a_{h+1}}(x),$$

where h is any positive integer. We shall prove the following theorem holds.

Theorem 1. *For any specific positive integer h and any integer $n \geq 0$, the following identity stands:*

$$\sum_{a_1 + a_2 + \cdots + a_{h+1} = n} B_{a_1}(x) B_{a_2}(x) \cdots B_{a_{h+1}}(x)$$

$$= \frac{1}{2^h \cdot h!} \cdot \sum_{j=1}^{h} \frac{M(h,j)}{(3x)^{2h-j}} \sum_{i=0}^{n} \frac{(n-i+j)!}{(n-i)!} \cdot \frac{B_{n-i+j}(x)}{(3x)^i} \cdot \binom{2h+i-j-1}{i},$$

where $M(h,i)$ is defined by $M(h,0) = 0$, $M(h,i) = \frac{(2h-i-1)!}{2^{h-i} \cdot (h-i)! \cdot (i-1)!}$ for all positive integers $1 \leq i \leq h$.

In particular, for $n = 0$, the following corollary can be deduced.

Corollary 1. *For any positive integer $h \geq 1$, the following formula holds:*

$$\sum_{j=1}^{h} M(h,j) \cdot j! \cdot (3x)^j \cdot B_j(x) = 2^h \cdot h! \cdot (3x)^{2h}.$$

The formula in Corollary 1 shows the close relationship among the balancing polynomials. For $h = 2$, the following corollary can be inferred by Theorem 1.

Corollary 2. *For any integer $n \geq 0$, we obtain*

$$\sum_{a+b+c=n} B_a(x) \cdot B_b(x) \cdot B_c(x) = \frac{1}{216x^3} \sum_{i=0}^{n} (n-i+1)(i+1)(i+2) \cdot \frac{B_{n-i+2}}{(3x)^i}$$

$$+ \frac{1}{72x^2} \sum_{i=0}^{n} (n-i+1)(n-i+2)(i+1) \cdot \frac{B_{n-i+3}}{(3x)^i}.$$

For $x = 1$, $h = 2$ *and* 3, according to Theorem 1 we can also infer the following corollaries:

Corollary 3. *For any integer $n \geq 0$, we obtain*

$$\sum_{a+b+c=n} B_{a+1} \cdot B_{b+1} \cdot B_{c+1} = \frac{1}{216} \sum_{i=0}^{n} (n-i+1)(i+1)(i+2) \cdot \frac{B_{n-i+2}}{3^i}$$

$$+ \frac{1}{72} \sum_{i=0}^{n} (n-i+1)(n-i+2)(i+1) \cdot \frac{B_{n-i+3}}{3^i}.$$

Corollary 4. *For any integer $n \geq 0$, we obtain:*

$$\sum_{a+b+c+d=n} B_{a+1} \cdot B_{b+1} \cdot B_{c+1} \cdot B_{d+1}$$

$$= \frac{1}{3888} \sum_{i=0}^{n} (n-i+1)(i+1)(i+2)(i+3)(i+4) \cdot \frac{B_{n-i+2}}{3^i}$$

$$+ \frac{1}{1296} \sum_{i=0}^{n} (n-i+1)(n-i+2)(i+1)(i+2)(i+3) \cdot \frac{B_{n-i+3}}{3^i}$$

$$+ \frac{1}{1296} \sum_{i=0}^{n} (n-i+1)(n-i+2)(n-i+3)(i+1)(i+2) \cdot \frac{B_{n-i+4}}{3^i}.$$

Corollary 5. *For any odd prime p, we have the congruence $M(p,i) \equiv 0 (mod\, p), 0 \leq i \leq p-1$.*

Corollary 6. *The balancing polynomials are essentially Chebyshev polynomials of the second kind, specifically $B_n(x) = U_n(3x)$. Taking $x = \frac{1}{3}x$ in Theorem 1, we can get the following:*

$$\sum_{a_1+a_2+\cdots+a_{h+1}=n} U_{a_1}(x) U_{a_2}(x) \cdots U_{a_{h+1}}(x)$$

$$= \frac{1}{2^h \cdot h!} \cdot \sum_{j=1}^{h} \frac{(2h-j-1)!}{2^{h-j} \cdot (h-j)! \cdot (j-1)! \cdot x^{2h-j}} \sum_{i=0}^{n} \frac{(n-i+j)!}{(n-i)!} \cdot \frac{U_{n-i+j}(x)}{x^i} \cdot \binom{2h+i-j-1}{i}.$$

Compared with [8], we give a more precise result for $\sum_{a_1+a_2+\cdots+a_{h+1}=n} U_{a_1}(x) U_{a_2}(x) \cdots U_{a_{h+1}}(x)$ with the specific expressions of $M(h,i)$. This shows our novelty.

Here, we list the first several terms of $M(h,i)$ in Table 1 in order to demonstrate the properties of the sequence $M(h,i)$ clearly.

Table 1. Values of $M(h,i)$.

$M(h,i)$	$i=1$	$i=2$	$i=3$	$i=4$	$i=5$	$i=6$	$i=7$	$i=8$
$h=1$	1							
$h=2$	1	1						
$h=3$	3	3	1					
$h=4$	15	15	6	1				
$h=5$	105	105	45	10	1			
$h=6$	945	945	420	105	15	1		
$h=7$	10,395	10,395	4725	1260	210	21	1	
$h=8$	135,135	135,135	62,370	17,325	3150	378	28	1

2. Several Lemmas

For the sake of clarity, several lemmas that are necessary for proving our theorem will be given in this section.

Lemma 1. *For the sequence $M(n,i)$, the following identity holds for all $1 \leq i \leq n$:*

$$M(n,i) = \frac{(2n-i-1)!}{2^{n-i} \cdot (n-i)! \cdot (i-1)!}.$$

Proof. We present a straightforward proof of this lemma by using mathematical introduction. It is obvious that

$$M(1,1) = \frac{0!}{1 \cdot 0! \cdot 0!} = 1.$$

This means Lemma 1 is valid for $n = 1$. Without loss of generality, we assume that Lemma 1 holds for $1 \le n = h$ and all $1 \le i \le h$. Then, we have

$$M(h, i) = \frac{(2h - i - 1)!}{2^{h-i} \cdot (h-i)! \cdot (i-1)!},$$

$$M(h, i+1) = \frac{(2h - i - 2)!}{2^{h-i-1} \cdot (h-i-1)! \cdot i!}.$$

According to the definitions of $M(n, i)$, it is easy to find that

$$
\begin{aligned}
M(h+1, i+1) &= (2h - 1 - i) \cdot M(h, i+1) + M(h, i) \\
&= (2h - 1 - i) \cdot \frac{2(h-i)}{(2h-i-1)i} \cdot M(h,i) + M(h,i) \\
&= \frac{2h - i}{i} M(h,i) = \frac{(2h-i)!}{2^{h-i} \cdot (h-i)! \cdot i!} \\
&= \frac{(2(h+1) - (i+1) - 1)!}{2^{h-i} \cdot (h-i)! \cdot i!}.
\end{aligned}
$$

Thus, Lemma 1 is also valid for $n = h + 1$. From now on, Lemma 1 has been proved. $\square$

Lemma 2. *If we have a function $f(t) = \frac{1}{1 - 6xt + t^2}$, then for any positive integer n, real numbers x and t with $|t| < |3x|$, the following identity holds:*

$$2^n \cdot n! \cdot f^{n+1}(t) = \sum_{i=1}^{n} M(n, i) \cdot \frac{f^{(i)}(t)}{(3x - t)^{2n-i}},$$

where $f^{(i)}(t)$ denotes the i-th order derivative of $f(t)$, with respect to variable t and $M(n, i)$, which is defined in the theorem.

Proof. Similarly, Lemma 2 will be proved by mathematical induction. We start by showing that Lemma 2 is valid for $n = 1$. Using the properties of the derivative, we have:

$$f'(t) = (6x - 2t) \cdot f^2(t),$$

or

$$2f^2(t) = \frac{f'(t)}{3x - t} = M(1,1) \cdot \frac{f'(t)}{3x - t}.$$

This is in fact true and provides the main idea to show the following steps. Without loss of generality, we assume that Lemma 2 holds for $1 \le n = h$. Then, we have

$$2^h \cdot h! \cdot f^{h+1}(t) = \sum_{i=1}^{h} M(h, i) \cdot \frac{f^{(i)}(t)}{(3x - t)^{2h-i}}. \tag{2}$$

As an immediate consequence, we can tell by (2), the properties of $M(n, i)$, and the derivative, we get

$$2^h \cdot (h+1)! \cdot f^h(t) \cdot f'(t) = 2^{h+1} \cdot (h+1)! \cdot (3x - t) \cdot f^{h+2}(t)$$

$$= \sum_{i=1}^{h} \frac{M(h, i)}{(3x - t)^{2h-i}} \cdot f^{(i+1)}(t) + \sum_{i=1}^{h} \frac{(2h - i) M(h, i)}{(3x - t)^{2h-i+1}} \cdot f^{(i)}(t)$$

$$
\begin{aligned}
&= \frac{M(h,h)}{(3x-t)^h} \cdot f^{(h+1)}(t) + \sum_{i=1}^{h-1} \frac{M(h,i)}{(3x-t)^{2h-i}} \cdot f^{(i+1)}(t) + \frac{(2h-1)M(h,1)}{(3x-t)^{2h}} \cdot f'(t) \\
&\quad + \sum_{i=1}^{h-1} \frac{(2h-i-1)M(h,i+1)}{(3x-t)^{2h-i}} \cdot f^{(i+1)}(t) \\
&= \frac{M(h+1,h+1)}{(3x-t)^h} \cdot f^{(h+1)}(t) + \frac{M(h+1,1)}{(3x-t)^{2h}} \cdot f'(t) + \sum_{i=1}^{h-1} \frac{M(h+1,i+1)}{(3x-t)^{2h-i}} \cdot f^{(i+1)}(t) \\
&= \frac{M(h+1,h+1)}{(3x-t)^h} \cdot f^{(h+1)}(t) + \frac{M(h+1,1)}{(3x-t)^{2h}} \cdot f'(t) + \sum_{i=2}^{h} \frac{M(h+1,i)}{(3x-t)^{2h+1-i}} \cdot f^{(i)}(t) \\
&= \sum_{i=1}^{h+1} M(h+1,i) \cdot \frac{f^{(i)}(t)}{(3x-t)^{2h+1-i}}.
\end{aligned}
\tag{3}
$$

Then, it is deduced that

$$
2^{h+1} \cdot (h+1)! \cdot (3x-t) \cdot f^{h+2}(t) = \sum_{i=1}^{h+1} M(h+1,i) \cdot \frac{f^{(i)}(t)}{(3x-t)^{2h+1-i}},
$$

or

$$
2^{h+1} \cdot (h+1)! \cdot f^{h+2}(t) = \sum_{i=1}^{h+1} M(h+1,i) \cdot \frac{f^{(i)}(t)}{(3x-t)^{2h+2-i}}.
$$

Thus, Lemma 2 is also valid for $n = h+1$. From now on, Lemma 2 has been proved. $\square$

Lemma 3. *The following power series expansion holds for arbitrary positive integers h and k:*

$$
\frac{f^{(h)}(t)}{(3x-t)^k} = \frac{1}{(3x)^k} \sum_{n=0}^{\infty} \left(\sum_{i=0}^{n} \frac{(n-i+h)!}{(n-i)!} \cdot \frac{B_{n-i+h}(x)}{(3x)^i} \cdot \binom{i+k-1}{i} \right) t^n,
$$

where t and x are any real numbers with $|t| < |3x|$.

Proof. According to the definition of the balancing polynomials $B_n(x)$, we have:

$$
f(t) = \frac{1}{1-6xt+t^2} = \sum_{n=0}^{\infty} B_n(x) \cdot t^n.
$$

For any positive integer h, from the properties of the power series, we can obtain

$$
\begin{aligned}
f^{(h)}(t) &= \sum_{n=0}^{\infty} (n+h)(n+h-1)\cdots(n+1) \cdot B_{n+h}(x) \cdot t^n \\
&= \sum_{n=0}^{\infty} \frac{(n+h)!}{n!} \cdot B_{n+h}(x) \cdot t^n.
\end{aligned}
\tag{4}
$$

For all real t and x with $|t| < |3x|$, we have the following power series expansion:

$$
\frac{1}{3x-t} = \frac{1}{3x} \cdot \sum_{n=0}^{\infty} \frac{t^n}{(3x)^n},
$$

and

$$
\frac{1}{(3x-t)^k} = \frac{1}{(3x)^k} \cdot \sum_{n=0}^{\infty} \binom{n+k-1}{n} \cdot \frac{t^n}{(3x)^n},
\tag{5}
$$

with any positive integer k. Then, it is found that

$$
\frac{f^{(h)}(t)}{(3x-t)^k}
$$

$$
= \frac{1}{(3x)^k} \cdot \left(\sum_{n=0}^{\infty} \frac{(n+h)!}{n!} \cdot B_{n+h}(x) \cdot t^n \right) \left(\sum_{n=0}^{\infty} \binom{n+k-1}{n} \cdot \frac{t^n}{(3x)^n} \right)
$$

$$
= \frac{1}{(3x)^k} \sum_{n=0}^{\infty} \left(\sum_{i+j=n} \frac{(j+h)!}{j!} \cdot B_{j+h}(x) \cdot \binom{i+k-1}{i} \cdot \frac{1}{(3x)^i} \right) t^n
$$

$$
= \frac{1}{(3x)^k} \sum_{n=0}^{\infty} \left(\sum_{i=0}^{n} \frac{(n-i+h)!}{(n-i)!} \cdot B_{n-i+h}(x) \cdot \binom{i+k-1}{i} \cdot \frac{1}{(3x)^i} \right) t^n,
$$

where we have used the multiplicative of the power series. Lemma 3 has been proved. $\square$

3. Proof of Theorem

Based on the lemmas in the above section, it is easy to deduce the proof of Theorem 1. For any positive integer h, we can derive

$$
2^h \cdot h! \cdot f^{h+1}(t) = 2^h \cdot h! \cdot \left(\sum_{n=0}^{\infty} B_n(x) \cdot t^n \right)^{h+1}
$$

$$
= 2^h \cdot h! \cdot \sum_{n=0}^{\infty} \left(\sum_{a_1+a_2+\cdots+a_{h+1}=n} B_{a_1}(x) B_{a_2}(x) \cdots B_{a_{h+1}}(x) \right) \cdot t^n. \tag{6}
$$

On the other hand, by the observation made in Lemma 3, it is deduced that

$$
2^h \cdot h! \cdot f^{h+1}(t) = \sum_{j=1}^{h} M(h,j) \cdot \frac{f^{(j)}(t)}{(3x-t)^{2h-j}}
$$

$$
= \sum_{j=1}^{h} \frac{M(h,j)}{(3x)^{2h-j}} \cdot \left(\sum_{n=0}^{\infty} \left(\sum_{i=0}^{n} \frac{(n-i+j)!}{(n-i)!} \cdot B_{n-i+j}(x) \cdot \binom{2h+i-j-1}{i} \cdot \frac{1}{(3x)^i} \right) t^n \right)
$$

$$
= \sum_{n=0}^{\infty} \left(\sum_{j=1}^{h} \frac{M(h,j)}{(3x)^{2h-j}} \sum_{i=0}^{n} \frac{(n-i+j)!}{(n-i)!} \cdot \frac{B_{n-i+j}(x)}{(3x)^i} \cdot \binom{2h+i-j-1}{i} \right) \cdot t^n. \tag{7}
$$

Altogether, we obtain the identity:

$$
2^h \cdot h! \sum_{a_1+a_2+\cdots+a_{h+1}=n} B_{a_1}(x) B_{a_2}(x) \cdots B_{a_{h+1}}(x)
$$

$$
= \sum_{j=1}^{h} \frac{M(h,j)}{(3x)^{2h-j}} \sum_{i=0}^{n} \frac{(n-i+j)!}{(n-i)!} \cdot \frac{B_{n-i+j}(x)}{(3x)^i} \cdot \binom{2h+i-j-1}{i}.
$$

This proves Theorem 1.

4. Conclusions

In this paper, a representation of a linear combination of balancing polynomials $B_i(x)$ (see Theorem 1) is obtained. Moreover, the specific expressions of $M(h,i)$ is given by using mathematical induction (see Lemma 1).

Theorem 1 can be reduced to various studies for the specific values of x, n, and h in the literature. For example, if $n=0$, our results reduce to Corollary 1. Taking $h=2$, our results reduce to Corollary 2. Taking $x=1$, $h=2,3$, our results reduce to Corollary 3 and Corollary 4, respectively.

Funding: This work is supported by the N. S. F. (11771351) of China.

Acknowledgments: The author would like to thank the Editor and the referees for their very helpful and detailed comments, which have significantly improved the presentation of this paper.

Conflicts of Interest: The author declares that there are no conflicts of interest regarding the publication of this paper.

References

1. Behera, A.; Panda, G.K. On the square roots of triangular numbers. *Fibonacci Quart.* **1999**, *37*, 98–105.
2. Panda, G.K.; Komatsu, T.; Davala, R.K. Reciprocal sums of sequences involving balancing and lucas-balancing numbers. *Math. Rep.* **2018**, *20*, 201–214.
3. Panda, G.K. Some fascinating properties of balancing numbers. *Congr. Numer.* **2009**, *194*, 185–189.
4. Patel, B.K.; Ray, P.K. The period, rank and order of the sequence of balancing numbers modulo m. *Math. Rep.* **2016**, *18*, 395–401.
5. Komatsu, T.; Szalay, L. Balancing with binomial coefficients. *Int. J. Number. Theory* **2014**, *10*, 1729–1742. [CrossRef]
6. Finkelstein, R. The house problem. *Am. Math. Mon.* **1965**, *72*, 1082–1088. [CrossRef]
7. Ray, P.K. Balancing and Cobalancing Numbers. Ph.D. Thesis, National Institute of Technology, Rourkela, India, 2009.
8. Zhang, Y.X.; Chen, Z.Y. A new identity involving the Chebyshev polynomials. *Mathematics* **2018**, *6*, 244. [CrossRef]
9. Zhao, J.H.; Chen, Z.Y. Some symmetric identities involving Fubini polynomials and Euler numbers. *Symmetry* **2018**, *10*, 303.
10. Ma, Y.K.; Zhang, W.P. Some identities involving Fibonacci polynomials and Fibonacci numbers. *Mathematics* **2018**, *6*, 334. [CrossRef]
11. Kim, D.S.; Kim, T. A generalization of power and alternating power sums to any Appell polynomials. *Filomat* **2017**, *1*, 141–157. [CrossRef]

Article

Note on Type 2 Degenerate q-Bernoulli Polynomials

Dae San Kim [1], Dmitry V. Dolgy [2], Jongkyum Kwon [3,*] and Taekyun Kim [4]

1   Department of Mathematics, Sogang University, Seoul 121-742, Korea
2   Kwangwoon Institute for Advanced Studies, Kwangwoon University, Seoul 139-701, Korea
3   Department of Mathematics Education and ERI, Gyeongsang National University, Jinju, Gyeongsangnamdo 52828, Korea
4   Department of Mathematics, Kwangwoon University, Seoul 139-701, Korea
*   Correspondence: mathkjk26@gnu.ac.kr

Received: 17 June 2019; Accepted: 12 July 2019; Published: 13 July 2019

Abstract: The purpose of this paper is to introduce and study type 2 degenerate q-Bernoulli polynomials and numbers by virtue of the bosonic p-adic q-integrals. The obtained results are, among other things, several expressions for those polynomials, identities involving those numbers, identities regarding Carlitz's q-Bernoulli numbers, identities concerning degenerate q-Bernoulli numbers, and the representations of the fully degenerate type 2 Bernoulli numbers in terms of moments of certain random variables, created from random variables with Laplace distributions. It is expected that, as was done in the case of type 2 degenerate Bernoulli polynomials and numbers, we will be able to find some identities of symmetry for those polynomials and numbers.

Keywords: type 2 degenerate q-Bernoulli polynomials; p-adic q-integral

MSC: 11S80; 11B83

1. Introduction

There are various ways of studying special polynomials and numbers, including generating functions, combinatorial methods, umbral calculus techniques, matrix theory, probability theory, p-adic analysis, differential equations, and so on.

In [1], it was shown that odd integer power sums (alternating odd integer power sums) can be represented in terms of some values of the type 2 Bernoulli polynomials (the type 2 Euler polynomials). In addition, some identities of symmetry, involving the type 2 Bernoulli polynomials, odd integer power sums, the type 2 Euler polynomials, and alternating odd integer power sums, were obtained by introducing appropriate quotients of bosonic and fermionic p-adic integrals on $\mathbb{Z}_p$. Furthermore, in [1], it was shown that the moments of two random variables, constructed from random variables with Laplace distributions, are closely connected with the type 2 Bernoulli numbers and the type 2 Euler numbers.

In recent years, studying degenerate versions of various special polynomials and numbers, which began with the paper by Carlitz in [2], has attracted the interest of many mathematicians. For example, in [3], the degenerate type 2 Bernoulli and Euler polynomials, and their corresponding numbers were introduced and some properties of them, which include distribution relations, Witt type formulas, and analogues for the interpretation of integer power sums in terms of Bernoulli polynomials, were investigated by means of both types of p-adic integrals.

As a q-analogue of the Volkenborn integrals for uniformly differentiable functions, the bosonic p-adic q-integrals were introduced in [4] by Kim. These integrals, together with the fermionic p-adic integrals and the fermionic p-adic q-integrals, have proven to be very useful tools in studying many problems arising from number theory and combinatorics. For instance, in [5], the type 2 q-Bernoulli

(q-Euler) polynomials were introduced by virtue of the bosonic (fermionic) p-adic q-integrals. Then, it was noted, among other things, that the odd q-integer (alternating odd q-integer) power sums are expressed in terms of the type 2 q-Bernoulli (q-Euler) polynomials.

In this short paper, we would like to introduce the type 2 degenerate q-Bernoulli polynomials and the corresponding numbers by making use of the bosonic p-adic q-integrals, as a degenerate version of and also as a q-analogue of the type 2 Bernoulli polynomials, and derive several basic results for them. The obtained results are several expressions for those polynomials, identities involving those numbers, identities regarding Carlitz's q-Bernoulli numbers, identities concerning degenerate q-Bernoulli numbers, and the representations of the fully degenerate type 2 Bernoulli numbers ($q = 1$ and $x = 1$ cases of the type 2 degenerate q-Bernoulli polynomials) in terms of moments of certain random variables, created from random variables with Laplace distributions.

The motivation for introducing the type 2 degenerate q-Bernoulli polynomials and numbers is to study their number-theoretic and combinatorial properties, and their applications in mathematics and other sciences in general. One novelty of this paper is that they arise naturally by means of the bosonic p-adic q-integrals so that it is possible to easily find some identities of symmetry for those polynomials and numbers, as it was done, for example, in [1]. In the rest of this section, we recall what is needed in the latter part of the paper.

Throughout this paper, p is a fixed odd prime number. We use the standard notations $\mathbb{Z}_p$, $\mathbb{Q}_p$, and $\mathbb{C}_p$ to denote the ring of p-adic integers, the field of p-adic rational numbers, and the completion of the algebraic closure of $\mathbb{Q}_p$, respectively. The p-adic norm on $\mathbb{C}_p$ is normalized as $|p|_p = \frac{1}{p}$.

As is well known, the Bernoulli numbers are given by the recurrence relation

$$B_0 = 1, \quad (B+1)^n - \begin{cases} 1, & \text{if } n = 1, \\ 0, & \text{if } n > 1, \end{cases} \tag{1}$$

where, as usual, B^n are to be replaced by B_n (see [2,6,7]).

Additionally, the Bernoulli polynomials of degree n are given by

$$B_n(x) = \sum_{l=0}^{n} \binom{n}{l} B_l x^{n-l}, \tag{2}$$

(see [3,8,9]).

Let q be an indeterminate in $\mathbb{C}_p$. For $q \in \mathbb{C}_p$, we assume that $|1 - q|_p < p^{-\frac{1}{p-1}}$.

In [7], Carlitz considered the q-Bernoulli numbers which are given by the recurrence relation:

$$\beta_{0,q} = 1, \quad q(q\beta_q + 1)^n - \beta_{n,q} = \begin{cases} 1, & \text{if } n = 1, \\ 0, & \text{if } n > 1, \end{cases} \tag{3}$$

where β_q^n are to be replaced by $\beta_{n,q}$, as usual.

In addition, he defined the q-Bernoulli polynomials as

$$\beta_{n,q}(x) = \sum_{l=0}^{n} \binom{n}{l} [x]_q^{n-l} q^{lx} \beta_{l,q}, \quad (n \geq 0), \tag{4}$$

where $[x]_q = \frac{1-q^x}{1-q}$, (see [7]).

Recently, the type 2 Bernoulli polynomials have been defined as

$$\frac{t}{e^t - e^{-t}} e^{xt} = \sum_{n=0}^{\infty} b_n(x) \frac{t^n}{n!}, \tag{5}$$

(see [1,3,8]).

When $x = 0$, $b_n = b_n(0)$ are called the type 2 Bernoulli numbers.

From (5), we note that

$$\sum_{l=0}^{n-1}(2l+1)^k = \frac{1}{k+1}(b_{k+1}(2n)-b_{k+1}), \quad (k \geq 0).$$ (6)

Let f be a uniformly differentiable function on $\mathbb{Z}_p$. Then, Kim defined the p-adic q-integral of f on $\mathbb{Z}_p$ as

$$
\begin{aligned}
I_q(f) &= \int_{\mathbb{Z}_p} f(x)d\mu_q(x) \\
&= \lim_{N\to\infty} \frac{1}{[p^N]_q} \sum_{x=0}^{p^N-1} f(x)q^x \\
&= \lim_{N\to\infty} \sum_{x=0}^{p^N-1} f(x)\mu_q(x+p^N\mathbb{Z}_p),
\end{aligned}
$$ (7)

(see [4]). Here, we note that $\mu_q(x+p^N\mathbb{Z}_p) = \frac{q^x}{[p^N]_q}$ is a distribution but not a measure. The details on the existence of the p-adic q-integrals for uniformly differentiable functions f on $\mathbb{Z}_p$ can be found in [4,10].

From (7), we note that

$$qI_q(f_1) = I_q(f) + (q-1)f(0) + \frac{q-1}{\log q}f'(0),$$ (8)

where $f_1(x) = f(x+1)$.

By virtue of (8) and induction, we get

$$q^nI_q(f_n) = I_q(f) + (q-1)\sum_{l=0}^{n-1}q^l f(l) + \frac{q-1}{\log q}\sum_{l=0}^{n-1}q^l f'(l),$$ (9)

where $f_n(x) = f(x+n), \quad (n \geq 1)$.

The degenerate exponential function is defined by

$$e_\lambda^x(t) = (1+\lambda t)^{\frac{x}{\lambda}},$$ (10)

(see [11]), where $\lambda \in \mathbb{C}_p$ with $|\lambda|_p < p^{-\frac{1}{p-1}}$.

For brevity, we also set

$$e_\lambda(t) = e_\lambda^1(t) = (1+\lambda t)^{\frac{1}{\lambda}}.$$ (11)

Carlitz defined the degenerate Bernoulli polynomials as

$$\frac{t}{e_\lambda(t)-1}e_\lambda^x(t) = \sum_{n=0}^{\infty} B_{n,\lambda}(x)\frac{t^n}{n!},$$ (12)

where $B_{n,\lambda} = B_{n,\lambda}(0)$ are called the degenerate Bernoulli numbers.

From (12), we note that

$$\sum_{l=0}^{n-1}(l)_{k,\lambda} = \frac{1}{k+1}(B_{k+1,\lambda}(n)-B_{k+1,\lambda}), \quad (n \geq 0),$$ (13)

where $(l)_{0,\lambda} = 1, (l)_{k,\lambda} = l(l-\lambda)\cdots(l-(k-1)\lambda), \quad (k \geq 1)$.

In the special case of $\lambda = 1$, the falling factorial sequence (also called the Pochammer symbol) is given by

$$(l)_0 = 1, (l)_k = l(l-1) \cdots (l-(k-1)), \quad (k \geq 1). \tag{14}$$

In this paper, we study type 2 degenerate q-Bernoulli polynomials and investigate some identities and properties for these polynomials.

2. Type 2 Degenerate q-Bernoulli Polynomials

Throughout this section, we assume that $q \in \mathbb{C}_p$ with $|1-q|_p < p^{-\frac{1}{p-1}}$ and $\lambda \in \mathbb{C}_p$. Now, we define the *type 2 degenerate q-Bernoulli polynomials* by

$$\sum_{n=0}^{\infty} b_{n,q}(x \mid \lambda)\frac{t^n}{n!} = \frac{1}{2} \int_{\mathbb{Z}_p} e_{\lambda}^{[x+2y]_q}(t)d\mu_q(y). \tag{15}$$

By (15), we get

$$\frac{1}{2} \int_{\mathbb{Z}_p} \left([x+2y]_q \right)_{n,\lambda} d\mu_q(y) = b_{n,q}(x \mid \lambda), \quad (n \geq 0). \tag{16}$$

When $x = 1$, $b_{n,q}(\lambda) = b_{n,q}(1 \mid \lambda)$ are called the type 2 degenerate q-Bernoulli numbers. We observe here that

$$\lim_{q \to 1}\lim_{\lambda \to 0} \frac{1}{2} \int_{\mathbb{Z}_p} \left([x+2y]_q \right)_{n,\lambda} d\mu_q(y)$$
$$= \lim_{q \to 1}\lim_{\lambda \to 0} \frac{1}{2}b_{n,q}(x \mid \lambda) = b_n(x-1), \quad (n \geq 0). \tag{17}$$

The degenerate Stirling numbers of the first kind appear as the coefficients in the expansion

$$(x)_{n,\lambda} = \sum_{l=0}^{n} S_{1,\lambda}(n,l)x^l, \quad (n \geq 0), \tag{18}$$

(see [12]).

Thus, by (18), we get

$$\frac{1}{2} \int_{\mathbb{Z}_p} \left([x+2y]_q \right)_{n,\lambda} d\mu_q(y)$$
$$= \frac{1}{2} \sum_{l=0}^{n} S_{1,\lambda}(n,l) \int_{\mathbb{Z}_p} [x+2y]_q^l d\mu_q(y) \tag{19}$$
$$= \sum_{l=0}^{n} S_{1,\lambda}(n,l)b_{l,q}(x),$$

where $b_{l,q}(x)$ is the type 2 q-Bernoulli polynomials given by $\frac{1}{2} \int_{\mathbb{Z}_p} [x+2y]_q^n d\mu_q(y) = b_{n,q}(x), (n \geq 0)$, (see [5]).

Therefore, by (16) and (19), we obtain the following theorem.

Theorem 1. *For $n \geq 0$, we have*

$$b_{n,q}(x \mid \lambda) = \sum_{l=0}^{n} S_{1,\lambda}(n,l)b_{l,q}(x). \tag{20}$$

Now, we observe that

$$\frac{1}{2} \int_{\mathbb{Z}_p} \left([x + 2y]_q \right)_{n,\lambda} d\mu_q(y)$$

$$= \frac{1}{2} \sum_{l=0}^{n} S_{1,\lambda}(n,l) \int_{\mathbb{Z}_p} [x + 2y]_q^l d\mu_q(y) \tag{21}$$

$$= \frac{1}{2} \sum_{l=0}^{n} S_{1,\lambda}(n,l) \frac{1}{(1-q)^l} \sum_{m=0}^{l} \binom{l}{m} (-q^x)^m \frac{2m+1}{[2m+1]_q}.$$

Therefore, by (21), we obtain the following theorem.

Theorem 2. *For $n \geq 0$, we have*

$$b_{n,q}(x \mid \lambda) = \frac{1}{2} \sum_{l=0}^{n} \frac{S_{1,\lambda}(n,l)}{(1-q)^l} \sum_{m=0}^{l} \binom{l}{m} (-q^x)^m \frac{2m+1}{[2m+1]_q}. \tag{22}$$

From (16), we note that

$$\sum_{n=0}^{\infty} b_{n,q}(x \mid \lambda) \frac{t^n}{n!} = \frac{1}{2} \sum_{n=0}^{\infty} \int_{\mathbb{Z}_p} \left([x + 2y]_q \right)_{n,\lambda} d\mu_q(y) \frac{t^n}{n!}$$

$$= \frac{1}{2} \int_{\mathbb{Z}_p} \left(1 + \lambda t \right)^{\frac{[x+2y]_q}{\lambda}} d\mu_q(y) \tag{23}$$

$$= \sum_{n=0}^{\infty} \left(\sum_{k=0}^{n} \sum_{m=0}^{k} \binom{k}{m} q^{mx} [x]_q^{k-m} S_1(n,k) \lambda^{n-k} b_{m,q} \right) \frac{t^n}{n!},$$

where $S_1(n,k)$ are the Stirling numbers of the first kind and $b_{n,q}$ are the type 2 q-Bernoulli numbers.
Therefore, by (23), we get the following theorem.

Theorem 3. *For $n \geq 0$, we have*

$$b_{n,q}(x \mid \lambda) = \sum_{k=0}^{n} \sum_{m=0}^{k} \binom{k}{m} q^{mx} [x]_q^{k-m} S_1(n,k) \lambda^{n-k} b_{m,q}. \tag{24}$$

In particular,

$$b_{n,q}(\lambda) = \sum_{k=0}^{n} q^{kx} S_1(n,k) \lambda^{n-k} b_{k,q}.$$

In [4], Kim expressed Carlitz's q-Bernoulli polynomials in terms of the following p-adic q-integrals on $\mathbb{Z}_p$:

$$\int_{\mathbb{Z}_p} [x + y]_q^n d\mu_q(y) = \beta_{n,q}(x), \quad (n \geq 0). \tag{25}$$

From (9) and (25), we have

$$
\begin{aligned}
q^n \beta_{m,q}(n) &= \int_{\mathbb{Z}_p} q^n [x+n]_q^m d\mu_q(x) \\
&= \int_{\mathbb{Z}_p} [x]_q^m d\mu_q(x) + (q-1) \sum_{l=0}^{n-1} q^l [l]_q^m + m \sum_{l=0}^{n-1} [l]_q^{m-1} q^{2l} \\
&= \beta_{m,q} + (q-1) \sum_{l=0}^{n-1} q^l [l]_q^m + m \sum_{l=0}^{n-1} [l]_q^{m-1} q^{2l} \\
&= \beta_{m,q} + (m+1) \sum_{l=0}^{n-1} q^{2l} [l]_q^{m-1} - \sum_{l=0}^{n-1} q^l [l]_q^{m-1},
\end{aligned}
\tag{26}
$$

where n is a positive integer.

Therefore, we obtain the following theorem.

Theorem 4. *For $n \geq 0$, we have*

$$
q^n \beta_{m,q}(n) - \beta_{m,q} = (m+1) \sum_{l=0}^{n-1} q^{2l} [l]_q^{m-1} - \sum_{l=0}^{n-1} q^l [l]_q^{m-1}.
\tag{27}
$$

Let us take $f(x) = \left([x]_q \right)_{m,\lambda}$, $(m \geq 1)$. From (9), we have

$$
\begin{aligned}
q^n &\int_{\mathbb{Z}_p} \left([x+n]_q \right)_{m,\lambda} d\mu_q(x) \\
&= \int_{\mathbb{Z}_p} \left([x]_q \right)_{m,\lambda} d\mu_q(x) + (q-1) \sum_{l=0}^{n-1} q^l \left([l]_q \right)_{m,\lambda} \\
&\quad + \sum_{l=0}^{n-1} \left(\sum_{k=0}^{m-1} \frac{1}{[l]_q - k\lambda} \right) \left([l]_q \right)_{m,\lambda} q^{2l}.
\end{aligned}
\tag{28}
$$

In [13], the degenerate q-Bernoulli polynomials are defined by Kim as

$$
\int_{\mathbb{Z}_p} e_\lambda^{[x+y]_q}(t) d\mu_q(y) = \sum_{n=0}^{\infty} \beta_{n,\lambda,q}(x) \frac{t^n}{n!}.
\tag{29}
$$

In particular, the degenerate q-Bernoulli numbers are given by $\beta_{n,\lambda,q} = \beta_{n,\lambda,q}(0)$.

From (29), we have

$$
\int_{\mathbb{Z}_p} \left([x+y]_q \right)_{n,\lambda} d\mu_q(y) = \beta_{n,\lambda,q}(x), \quad (n \geq 0).
\tag{30}
$$

By (28) and (30), this completes the proof for the next theorem.

Theorem 5. *For $m, n \in \mathbb{N}$, we have*

$$
\begin{aligned}
q^n &\beta_{m,\lambda,q}(n) - \beta_{m,\lambda,q} \\
&= (q-1) \sum_{l=0}^{n-1} q^l \left([l]_q \right)_{m,\lambda} + \sum_{l=0}^{n-1} \left(\sum_{k=0}^{m-1} \frac{1}{[l]_q - k\lambda} \right) \left([l]_q \right)_{m,\lambda} q^{2l}.
\end{aligned}
\tag{31}
$$

Let us take $f(x) = \left([2x+1]_q\right)_{m,\lambda}$, $(m \geq 1)$. From (9), we have

$$q^n \int_{\mathbb{Z}_p} \left([2x+2n+1]_q\right)_{m,\lambda} d\mu_q(x)$$

$$= \int_{\mathbb{Z}_p} \left([2x+1]_q\right)_{m,\lambda} d\mu_q(x) + (q-1)\sum_{l=0}^{n-1} q^l \left([2l+1]_q\right)_{m,\lambda} \tag{32}$$

$$+ 2\sum_{l=0}^{n-1}\left(\sum_{k=0}^{m-1} \frac{1}{[2l+1]_q - k\lambda}\right)\left([2l+1]_q\right)_{m,\lambda} q^{3l+1}.$$

From (16) and (32), we have

$$q^n b_{m,q}(2n+1 \mid \lambda) - b_{m,q}(\lambda)$$

$$= \frac{q-1}{2}\sum_{l=0}^{n-1} q^l \left([2l+1]_q\right)_{m,\lambda} + \sum_{l=0}^{n-1}\left(\sum_{k=0}^{m-1}\frac{1}{[2l+1]_q - k\lambda}\right)\left([2l+1]_q\right)_{m,\lambda} q^{3l+1}. \tag{33}$$

Therefore, by (33), we obtain the following theorem.

Theorem 6. *For $m, n \in \mathbb{N}$, we have*

$$\frac{q-1}{2}\sum_{l=0}^{n-1} q^l \left([2l+1]_q\right)_{m,\lambda} + \sum_{l=0}^{n-1}\left(\sum_{k=0}^{m-1}\frac{1}{[2l+1]_q - k\lambda}\right)\left([2l+1]_q\right)_{m,\lambda} q^{3l+1}$$

$$= q^n b_{m,q}(2n+1 \mid \lambda) - b_{m,q}(\lambda). \tag{34}$$

From (7), we can derive the following integral equation:

$$\int_{\mathbb{Z}_p} f(x) d\mu_q(x) = \lim_{N \to \infty} \frac{1}{[p^N]_q} \sum_{x=0}^{p^N-1} f(x) q^x$$

$$= \lim_{N \to \infty} \frac{1}{[dp^N]_q} \sum_{x=0}^{dp^N-1} f(x) q^x$$

$$= \lim_{N \to \infty} \frac{1}{[dp^N]_q} \sum_{a=0}^{d-1}\sum_{x=0}^{p^N-1} f(a+dx) q^{a+dx} \tag{35}$$

$$= \sum_{a=0}^{d-1} q^a \frac{1}{[d]_q} \lim_{N \to \infty} \frac{1}{[p^N]_{q^d}} \sum_{x=0}^{p^N-1} f(a+dx) q^{dx}$$

$$= \frac{1}{[d]_q}\sum_{a=0}^{d-1} q^a \int_{\mathbb{Z}_p} f(a+dx) d\mu_{q^d}(x),$$

where d is a positive integer.

Lemma 1. *For $d \in \mathbb{N}$, we have*

$$\int_{\mathbb{Z}_p} f(x) d\mu_q(x) = \frac{1}{[d]_q}\sum_{a=0}^{d-1} q^a \int_{\mathbb{Z}_p} f(a+dx) d\mu_{q^d}(x). \tag{36}$$

We obtain the following theorem from Lemma 1.

Theorem 7. *For $n, d \in \mathbb{N}$, we have*

$$b_{n,q}(\lambda) = [d]_q^{n-1} \sum_{a=0}^{d-1} q^a b_{n,q^d}\left(\frac{2a+1}{d} \mid \frac{\lambda}{[d]_q}\right).$$

(37)

Proof. Let us apply Lemma 1 with $f(x) = \left([2x+1]_q\right)_{n,\lambda}$, $(n \in \mathbb{N})$. Then, by virtue of (16), we have

$$\int_{\mathbb{Z}_p} \left([2x+1]_q\right)_{n,\lambda} d\mu_q(x) = \frac{1}{[d]_q} \sum_{a=0}^{d-1} q^a \int_{\mathbb{Z}_p} \left([2(a+dx)+1]_q\right)_{n,\lambda} d\mu_{q^d}(x)$$

$$= \frac{1}{[d]_q} \sum_{a=0}^{d-1} q^a [d]_q^n \int_{\mathbb{Z}_p} \left(\left[\frac{2a+1}{d} + 2x\right]_{q^d}\right)_{n, \frac{\lambda}{[d]_q}} d\mu_{q^d}(x)$$

$$= [d]_q^{n-1} \sum_{a=0}^{d-1} q^a \int_{\mathbb{Z}_p} \left(\left[\frac{2a+1}{d} + 2x\right]_{q^d}\right)_{n, \frac{\lambda}{[d]_q}} d\mu_{q^d}(x)$$

$$= 2[d]_q^{n-1} \sum_{a=0}^{d-1} q^a b_{n,q^d}\left(\frac{2a+1}{d} \mid \frac{\lambda}{[d]_q}\right).$$

$\square$

3. Further Remarks

Assume that $X_1, X_2, X_3, \cdots$ are independent random variables, each of which has the Laplace distribution with parameters 0 and 1. Namely, each of them has the probability density function given by $\frac{1}{2}\exp(-|x|)$.

Let Z be the random variable given by $Z = \sum_{k=1}^{\infty} \frac{X_k}{2k\pi}$. In addition, let b_n be the type 2 Bernoulli numbers defined by

$$\frac{t}{e^t - e^{-t}} = \sum_{n=0}^{\infty} b_n \frac{t^n}{n!}.$$

(38)

Then, it was shown in [1] that

$$\sum_{n=0}^{\infty} E[Z^n] \frac{(it)^n}{n!} = \frac{t}{e^{\frac{t}{2}} - e^{-\frac{t}{2}}}$$

$$= \sum_{n=0}^{\infty} \left(\frac{1}{2}\right)^{n-1} b_n \frac{t^n}{n!}.$$

(39)

Thereby, it was obtained that

$$i^n E[Z^n] = \left(\frac{1}{2}\right)^{n-1} b_n.$$

(40)

Before proceeding further, we recall that the Volkenborn integral (also called the p-adic invariant integral) for a uniformly differentiable function f on $\mathbb{Z}_p$ is given by

$$\int_{\mathbb{Z}_p} f(y) d\mu(y) = \lim_{N \to \infty} \frac{1}{p^N} \sum_{y=0}^{p^N - 1} f(y).$$

(41)

Then, it is well known (see [14]) that this integral satisfies the following integral equation:

$$\int_{\mathbb{Z}_p} f(y+1) d\mu(y) = \int_{\mathbb{Z}_p} f(y) d\mu(y) + f'(0).$$

(42)

When $q = 1$ and $x = 1$, by virtue of (41), (15) becomes

$$\sum_{n=0}^{\infty} b_n(\lambda)\frac{t^n}{n!} = \frac{1}{2}\int_{\mathbb{Z}_p} e_{\lambda}^{1+2y}(t)d\mu(y)$$
$$= \frac{\frac{1}{\lambda}\log(1+\lambda t)}{e_{\lambda}(t) - e_{\lambda}^{-1}(t)}. \tag{43}$$

Here, $b_n(\lambda)$ may be called the fully degenerate type 2 Bernoulli numbers, even though they were defined slightly differently in [3]. Replacing t with $\frac{2}{\lambda}\log(1+\lambda t)$ in (39) and by making use of (43), we have

$$\sum_{m=0}^{\infty} E[Z^m](2i)^m\frac{1}{m!}\left(\frac{\log(1+\lambda t)}{\lambda}\right)^m$$
$$= \sum_{n=0}^{\infty}\left(\sum_{m=0}^{n} S_{1,\lambda}(n,m)(2i)^m E[Z^m]\right)\frac{t^n}{n!} \tag{44}$$
$$= 2\sum_{n=0}^{\infty} b_n(\lambda)\frac{t^n}{n!}.$$

Here, $S_{1,\lambda}(n,k)$ are the degenerate Stirling numbers of the first kind (see [12]) either given by

$$\frac{1}{m}\left(\frac{\log(1+\lambda t)}{\lambda}\right)^m = \sum_{n=m}^{\infty} S_{1,\lambda}(n,m)\frac{t^n}{n!}, \tag{45}$$

or given by

$$(x)_{n,\lambda} = \sum_{m=0}^{n} S_{1,\lambda}(n,m)x^m = \sum_{m=0}^{n} S_1(n,m)\lambda^{n-m}x^m. \tag{46}$$

Thus, by (44), we have shown that

$$2b_n(\lambda) = \sum_{m=0}^{n} S_{1,\lambda}(n,m)(2i)^m E[Z^m].$$

Here, we remark that we only considered $q = 1$ and $x = 1$ cases of (15), namely the fully degenerate type 2 Bernoulli numbers. This is because we do not see how to express type 2 degenerate q-Bernoulli polynomials or type 2 degenerate q-Bernoulli numbers in terms of the moments of some suitable random variables, constructed from random variables with Laplace distributions. We leave this as an open problem to the interested reader.

4. Conclusions

Studies on various special polynomials and numbers have been preformed using several different methods, such as generating functions, combinatorial methods, umbral calculus techniques, matrix theory, probability theory, p-adic analysis, differential equations, and so on.

One way of introducing new special polynomials and numbers is to study various degenerate versions of some known special polynomials and numbers, which began with Carlitz's paper in [2]. Actually, degenerate versions were investigated not only for some polynomials but also for a transcendental function, namely the gamma function. For this, we refer the reader to [11]. Another way of introducing new special polynomials and numbers is to study various q-analogues of some known special polynomials and numbers. The bosonic p-adic q-integrals, together with the fermionic p-adic q-integrals, turned out to be very powerful and fruitful tools for naturally constructing such q-analogues. They were introduced by Kim in [4] and have been widely used ever since their invention.

In this paper, the type 2 degenerate q-Bernoulli polynomials and the corresponding numbers were introduced and investigated as a degenerate version of and also as a q-analogue of type 2 Bernoulli polynomials by making use of the bosonic p-adic q-integrals [1,3,5]. Here, as an introductory paper on the subject, only very basic results were obtained. The obtained results are several expressions for those polynomials, identities involving those numbers, identities regarding Carlitz's q-Bernoulli numbers, identities concerning degenerate q-Bernoulli numbers, and the representations of the fully degenerate type 2 Bernoulli numbers ($q = 1$ and $x = 1$ cases of the type 2 degenerate q-Bernoulli polynomials) in terms of moments of certain random variables, created from random variables with Laplace distributions. We are planning to study more detailed results relating to these polynomials and numbers in a forthcoming paper.

Author Contributions: All authors contributed equally to the manuscript and typed, read, and approved the final manuscript.

Funding: This work was supported by the National Research Foundation of Korea (NRF) grant funded by the Korea government (MEST) (No. 2017R1E1A1A03070882).

Acknowledgments: The authors would like to thank the referees for their valuable comments and suggestions.

Conflicts of Interest: The authors declare no conflict of interest.

References

1. Kim, D.S.; Kim, H.Y.; Kim, D.; Kim, T. Identities of symmetry for type 2 Bernoulli and Euler polynomials. *Symmetry* **2019**, *11*, 613. [CrossRef]
2. Carlitz, L. Degenerate Stirling, Bernoulli and Eulerian numbers. *Utilitas Math.* **1979**, *15*, 51–88.
3. Kim, D.S.; Kim, H.Y.; Pyo, S.-S.; Kim, T. Some identities of special numbers and polynomials arising from p-adic integrals on $\mathbb{Z}_p$. *Adv. Differ. Equ.* **2019**, *2019*, 190. [CrossRef]
4. Kim, T. q-Volkenborn integration. *Russ. J. Math. Phys.* **2002**, *9*, 288–299.
5. Kim, D.S.; Kim, T.; Kim, H.Y.; Kwon, J. A note on type 2 q-Bernoulli and type 2 q-Euler polynomials. *J. Inequal. Appl.* **2019**, accepted. [CrossRef]
6. Araci, S.; Acikgoz, M. A note on the Frobenius-Euler numbers and polynomials associated with Bernstein polynomials. *Adv. Stud. Contemp. Math.* **2012**, *22*, 399–406.
7. Carlitz, L. Expansions of q-Bernoulli numbers. *Duke Math. J.* **1958**, *25*, 355–364. [CrossRef]
8. Jang, G.-W.; Kim, T. A note on type 2 degenerate Euler and Bernoulli polynomials. *Adv. Stud. Contemp. Math.* **2019**, *29*, 147–159.
9. Kim, D.S.; Dolgy, D.V.; Kim, D.; Kim, T. Some identities on r-central factorial numbers and r-central Bell polynomials. *Adv. Differ. Equ.* **2019**, accepted. [CrossRef]
10. Diarra, B. The p-adic q-distributions, Advances in ultrametric analysis. *Contemp. Math.* **2013**, *596*, 45–62.
11. Kim, T.; Kim, D.S. Degenerate Laplace transform and degenerate gamma function. *Russ. J. Math. Phys.* **2017**, *24*, 241–248. [CrossRef]
12. Kim, D.S.; Kim, T.; Jang, G.-W. A note on degenerate Stirling numbers of the first kind. *Proc. Jangjeon Math. Soc.* **2018**, *21*, 393–404.
13. Kim, T. On degenerate q-Bernoulli polynomials. *Bull. Korean Math. Soc.* **2016**, *53*, 1149–1156. [CrossRef]
14. Schikhof, W.H. *Ultrametric Calculus. An Introduction to p-Adic Analysis*; Cambridge Studies in Advanced Mathematics, 4; Cambridge University Press: Cambridge, UK, 1984.

 *symmetry*

Article

Some Identities of Ordinary and Degenerate Bernoulli Numbers and Polynomials

Dmitry V. Dolgy [1], Dae San Kim [2], Jongkyum Kwon [3],* and Taekyun Kim [4]

[1] Hanrimwon, Kwangwoon University, Seoul 139-701, Korea
[2] Department of Mathematics, Sogang University, Seoul 121-742, Korea
[3] Department of Mathematics Education and ERI, Gyeongsang National University, Jinju, Gyeongsangnamdo 52828, Korea
[4] Department of Mathematics, Kwangwoon University, Seoul 139-701, Korea
*   Correspondence: mathkjk26@gnu.ac.kr

Received: 28 May 2019; Accepted: 26 June 2019; Published: 1 July 2019

Abstract: In this paper, we investigate some identities on Bernoulli numbers and polynomials and those on degenerate Bernoulli numbers and polynomials arising from certain p-adic invariant integrals on $\mathbb{Z}_p$. In particular, we derive various expressions for the polynomials associated with integer power sums, called integer power sum polynomials and also for their degenerate versions. Further, we compute the expectations of an infinite family of random variables which involve the degenerate Stirling polynomials of the second and some value of higher-order Bernoulli polynomials.

Keywords: Bernoulli polynomials; degenerate Bernoulli polynomials; random variables; p-adic invariant integral on $\mathbb{Z}_p$; integer power sums polynomials; Stirling polynomials of the second kind; degenerate Stirling polynomials of the second kind

---

1. Introduction

We begin this section by reviewing some known facts. In more detail, we recall the integral equation for the p-adic invariant integral of a uniformly differentiable function on $\mathbb{Z}_p$ and its generalizations, the expression in terms of some values of Bernoulli polynomials for the integer power sums, and the p-adic integral representaions of Bernoulli polynomials and of their generating functions.

Throughout this paper, $\mathbb{Z}_p$, $\mathbb{Q}_p$ and $\mathbb{C}_p$ will denote the ring of p-adic integers, the field of p-adic rational numbers and the completion of the algebraic closure of $\mathbb{Q}_p$, respectively. The p-adic norm is normalized as $|p|_p = \frac{1}{p}$. Let f be a uniformly differentiable function on $\mathbb{Z}_p$. Then the p-adic invariant integral of f (also called the Volkenborn integral of f) on $\mathbb{Z}_p$ is defined by

$$I_0(f) = \int_{\mathbb{Z}_p} f(x)d\mu_0(x) = \lim_{N \to \infty} \frac{1}{p^N} \sum_{x=0}^{p^N - 1} f(x)$$

$$= \lim_{N \to \infty} \sum_{x=0}^{p^N - 1} f(x)\mu_0(x + p^N \mathbb{Z}_p). \tag{1}$$

Here we note that $\mu_0(x + p^N \mathbb{Z}_p) = \frac{1}{p^N}$ is a distribution but not a measure. The existence of such integrals for uniformly differentiable functions on $\mathbb{Z}_p$ is detailed in [1,2]. It can be seen from (1) that

$$I_0(f_1) = I_0(f) + f'(0), \tag{2}$$

where $f_1(x) = f(x+1)$, and $f'(0) = \frac{df(x)}{dx}\big|_{x=0}$, (see [1,2]).

In general, by induction and with $f_n(x) = f(x+n)$, we can show that

$$I_0(f_n) = I_0(f) + \sum_{k=0}^{n-1} f'(k), \quad (n \in \mathbb{N}), \tag{3}$$

As is well known, the Bernoulli polynomials are given by the generating function (see [3–5])

$$\frac{t}{e^t - 1} e^{xt} = \sum_{n=0}^{\infty} B_n(x) \frac{t^n}{n!}, \tag{4}$$

When $x = 0$, $B_n = B_n(0)$ are called the Bernoulli numbers.
From (4), we note that (see [3–5])

$$B_n(x) = \sum_{l=0}^{n} \binom{n}{l} B_l x^{n-l}, \quad (n \geq 0), \tag{5}$$

and

$$B_0 = 1, \quad \sum_{k=0}^{n} \binom{n}{k} B_k - B_n = \begin{cases} 1, & \text{if } n = 1, \\ 0, & \text{if } n > 1, \end{cases}$$

Let (see [6–13])

$$S_p(n) = \sum_{k=1}^{n} k^p, \quad (n, p \in \mathbb{N}). \tag{6}$$

The generating function of $S_p(n)$ is given by

$$\sum_{p=0}^{\infty} S_p(n) \frac{t^p}{p!} = \sum_{k=1}^{n} e^{kt} = \frac{1}{t} \left(\frac{t}{e^t - 1} \left(e^{(n+1)t} - e^t \right) \right)$$

$$= \sum_{p=0}^{\infty} \left(\frac{B_{p+1}(n+1) - B_{p+1}(1)}{p+1} \right) \frac{t^p}{p!}. \tag{7}$$

Thus, by (7), we get

$$S_p(n) = \frac{B_{p+1}(n+1) - B_{p+1}(1)}{p+1}, \quad (n, p \in \mathbb{N}). \tag{8}$$

From (2), we have

$$\int_{\mathbb{Z}_p} e^{(x+y)t} d\mu_0(y) = \frac{t}{e^t - 1} e^{xt} = \sum_{n=0}^{\infty} B_n(x) \frac{t^n}{n!}. \tag{9}$$

By (9), we get (see [11,12])

$$\int_{\mathbb{Z}_p} (x+y)^n d\mu_0(y) = B_n(x), \quad (n \geq 0), \tag{10}$$

From (8) and (10), we can derive the following equation.

$$\int_{\mathbb{Z}_p} (x+k+1)^{p+1} d\mu_0(x) - \int_{\mathbb{Z}_p} x^{p+1} d\mu_0(x) = (p+1) \sum_{n=1}^{k} n^p, \quad (p \in \mathbb{N}). \tag{11}$$

Thus, by (6) and (11), and for $p \in \mathbb{N}$, we get

$$S_p(k) = \frac{1}{p+1} \left\{ \int_{\mathbb{Z}_p} (x+k+1)^{p+1} d\mu_0(x) - \int_{\mathbb{Z}_p} x^{p+1} d\mu_0(x) \right\}. \tag{12}$$

The purpose of this paper is to investigate some identities on Bernoulli numbers and polynomials and those on degenerate Bernoulli numbers and polynomials arising from certain p-adic invariant integrals on $\mathbb{Z}_p$.

The outline of this paper is as in the following. After reviewing well- known necessary results in Section 1, we will derive some identities on Bernoulli polynomials and numbers in Section 2. In particular, we will introduce the integer power sum polynomials and derive several expressions for them. In Section 3, we will obtain some identities on degenerate Bernoulli numbers and polynomials. Especially, we will introduce the degenerate integer power sum polynomials, a degenerate version of the integer power sum polynomials and deduce various representations of them. In the final Section 4, we will consider an infinite family of random variables and compute their expectations to see that they involve the degenerate Stirling polynomials of the second and some value of higher-order Bernoulli polynomials.

2. Some Identities of Bernoulli Numbers and Polynomials

For $p \in \mathbb{N}$, we observe that

$$
\begin{aligned}
(j+1)^{p+1} - j^{p+1} &= \sum_{i=0}^{p+1} \binom{p+1}{i} j^i - j^{p+1} \\
&= (p+1)j^p + \sum_{i=1}^{p-1} \binom{p+1}{i} j^i + 1.
\end{aligned}
\tag{13}
$$

Thus, we get

$$
(n+1)^{p+1} = \sum_{j=0}^{n} \left\{ (j+1)^{p+1} - j^{p+1} \right\} = (p+1) \sum_{j=0}^{n} j^p + \sum_{i=1}^{p-1} \binom{p+1}{i} \sum_{j=0}^{n} j^i + (n+1).
\tag{14}
$$

From (14), we have

$$
S_p(n) = \frac{1}{p+1} \left\{ (n+1)^{p+1} - (n+1) - \sum_{i=1}^{p-1} \binom{p+1}{i} S_i(n) \right\}.
\tag{15}
$$

Therefore, by (15), we obtain the following lemma.

Lemma 1. *For $n, p \in \mathbb{N}$, we have*

$$
\begin{aligned}
&\int_{\mathbb{Z}_p} (x+n+1)^{p+1} d\mu_0(x) - \int_{\mathbb{Z}_p} x^{p+1} d\mu_0(x) \\
=&(n+1)^{p+1} - (n+1) - \sum_{i=1}^{p-1} \binom{p+1}{i} \frac{1}{i+1} \\
&\times \left\{ \int_{\mathbb{Z}_p} (x+n+1)^{i+1} d\mu_0(x) - \int_{\mathbb{Z}_p} x^{i+1} d\mu_0(x) \right\}.
\end{aligned}
\tag{16}
$$

From Lemma 1, we note the following.

Corollary 1. *For $n, p \in \mathbb{N}$, we have*

$$
B_{p+1}(n+1) - B_{p+1} = (n+1)^{p+1} - (n+1) - \sum_{i=1}^{p-1} \binom{p+1}{i} \frac{1}{i+1} \left(B_{i+1}(n+1) - B_{i+1} \right).
\tag{17}
$$

For $n \in \mathbb{N}_0 = \mathbb{N} \cup \{0\}$, by (1), we get

$$\int_{\mathbb{Z}_p} \left(y + 1 - x \right)^n d\mu_0(y) = (-1)^n \int_{\mathbb{Z}_p} (y+x)^n d\mu_0(y). \tag{18}$$

From (18), we note that

$$B_n(1-x) = (-1)^n B_n(x), \quad (n \geq 0). \tag{19}$$

Now, we observe that, for $n \geq 1$,

$$\begin{aligned}
B_n(2) &= \sum_{l=0}^{n} \binom{n}{l} B_l(1) = B_0 + \binom{n}{1} B_1(1) + \sum_{l=2}^{n} \binom{n}{l} B_l(1) \\
&= B_0 + \binom{n}{1} B_1 + n + \sum_{l=2}^{n} \binom{n}{l} B_l = n + \sum_{l=0}^{n} \binom{n}{l} B_l \\
&= n + B_n(1).
\end{aligned} \tag{20}$$

Thus we have completed the proof for the next lemma.

Lemma 2. *For any $n \in \mathbb{N}_0$, the following identity is valid:*

$$B_n(2) = n + B_n + \delta_{n,1}, \tag{21}$$

where $\delta_{n,1}$ is the Kronecker's delta.

For any $n, m \in \mathbb{N}$ with $n, m \geq 2$, we have

$$\begin{aligned}
\int_{\mathbb{Z}_p} x^m (-1+x)^n d\mu_0(x) &= \sum_{i=0}^{n} \binom{n}{i} (-1)^{n-i} \int_{\mathbb{Z}_p} x^{m+i} d\mu_0(x) \\
&= \sum_{i=0}^{n} \binom{n}{i} (-1)^{n-i} B_{m+i} \\
&= (-1)^{n-m} \sum_{i=0}^{n} \binom{n}{i} B_{m+i}.
\end{aligned} \tag{22}$$

On the other hand,

$$\begin{aligned}
\int_{\mathbb{Z}_p} x^m (x-1)^n d\mu_0(x) &= \sum_{i=0}^{m} \binom{m}{i} \int_{\mathbb{Z}_p} (x-1)^{n+i} d\mu_0(x) \\
&= \sum_{i=0}^{m} \binom{m}{i} (-1)^{n+i} \int_{\mathbb{Z}_p} (x+2)^{n+i} d\mu_0(x) \\
&= \sum_{i=0}^{m} \binom{m}{i} (-1)^{n+i} \left(B_{n+i} + n + i \right) \\
&= \sum_{i=0}^{m} \binom{m}{i} (-1)^{n+i} B_{n+i} \\
&= \sum_{i=0}^{m} \binom{m}{i} B_{n+i}.
\end{aligned} \tag{23}$$

Therefore, by (22) and (23), we obtain the following theorem.

Theorem 1. *For any $m, n \in \mathbb{N}$ with $m, n \geq 2$, the following symmetric identity holds:*

$$(-1)^n \sum_{i=0}^{n} \binom{n}{i} B_{m+i} = (-1)^m \sum_{i=0}^{m} \binom{m}{i} B_{n+i}. \tag{24}$$

From (5), we note that

$$B_n(1) = \sum_{l=0}^{n} \binom{n}{l} B_l, \quad (n \geq 0).$$

For $n \geq 2$, we have

$$B_n = B_n(1) = \sum_{l=0}^{n} \binom{n}{l} B_l = \sum_{l=0}^{n} \binom{n}{l} B_{n-l}. \tag{25}$$

Now, we define the *integer power sum polynomials* by

$$S_p(n|x) = \sum_{k=0}^{n} (k+x)^p, \quad (n, p \in \mathbb{N}_0). \tag{26}$$

Note that $S_p(n|0) = S_p(n)$, $(n \in \mathbb{N}_0, p \in \mathbb{N})$.
For $N \in \mathbb{N}_0$, we have

$$t \sum_{k=0}^{N} e^{(k+x)t} = \int_{\mathbb{Z}_p} e^{(N+1+x+y)t} d\mu_0(y) - \int_{\mathbb{Z}_p} e^{(x+y)t} d\mu_0(y). \tag{27}$$

Then it is immediate to see from (27) that we have

$$\sum_{k=0}^{N} e^{(k+x)t} = \sum_{n=0}^{\infty} \frac{1}{n+1} \left\{ \int_{\mathbb{Z}_p} (N+1+x+y)^{n+1} d\mu_0(y) - \int_{\mathbb{Z}_p} (x+y)^{n+1} d\mu_0(y) \right\} \frac{t^n}{n!}. \tag{28}$$

Now, we see that (28) is equivalent to the next theorem.

Theorem 2. *For $n, N \in \mathbb{N}_0$, we have*

$$S_n(N|x) = \frac{1}{n+1} \left\{ B_{n+1}(x+N+1) - B_{n+1}(x) \right\}. \tag{29}$$

Let $\triangle$ denote the difference operator given by

$$\triangle f(x) = f(x+1) - f(x). \tag{30}$$

Then, by (30) and induction, we get

$$\triangle^n f(x) = \sum_{k=0}^{n} \binom{n}{k} (-1)^{n-k} f(x+k), \quad (n \geq 0). \tag{31}$$

Now, we can deduce the Equation (32) from (27) as in the following:

$$\sum_{k=0}^{N} e^{(k+x)t} = \frac{1}{t} e^{xt} \left(e^{(N+1)t} - 1 \right) \int_{\mathbb{Z}_p} e^{yt} d\mu_0(y)$$

$$= \frac{1}{e^t - 1} \left(\sum_{m=0}^{N+1} \binom{N+1}{m} (e^t - 1)^m - 1 \right) e^{xt}$$

$$= \frac{1}{e^t - 1} \sum_{m=1}^{N+1} \binom{N+1}{m} (e^t - 1)^m e^{xt}$$

$$= \sum_{m=0}^{N} \binom{N+1}{m+1} (e^t - 1)^m e^{xt}$$

$$= \sum_{n=0}^{\infty} \left\{ \sum_{m=0}^{N} \binom{N+1}{m+1} \sum_{k=0}^{m} \binom{m}{k} (-1)^{m-k} (k+x)^n \right\} \frac{t^n}{n!}$$

$$= \sum_{n=0}^{\infty} \left\{ \sum_{k=0}^{N} \sum_{m=k}^{N} \binom{N+1}{m+1} \binom{m}{k} (-1)^{m-k} (k+x)^n \right\} \frac{t^n}{n!}.$$

$$(32)$$

Therefore, (31) and (32) together yield the next theorem.

Theorem 3. *For $n, N \geq 0$, we have*

$$S_n(N|x) = \sum_{m=0}^{N} \binom{N+1}{m+1} \Delta^m x^n = \sum_{k=0}^{N} (k+x)^n T(N,k),$$
$$(33)$$

where $T(N,k) = \sum_{m=k}^{N} \binom{N+1}{m+1} \binom{m}{k} (-1)^{m-k}$.
In particular, we have

$$S_0(N|x) = \sum_{k=0}^{N} T(N,k) = N+1.$$

We recall here that the Stirling polynomials of the second kind $S_2(n,k|x)$ are given by (see [14])

$$\frac{1}{k!} (e^t - 1)^k e^{xt} = \sum_{n=k}^{\infty} S_2(n,k|x) \frac{t^n}{n!}.$$
$$(34)$$

Note here that $S_2(n,k|0) = S_2(n,k)$ are Stirling numbers of the second kind. Then, we can show that, for integers $n, m \geq 0$, we have

$$\frac{1}{m!} \Delta^m x^n = \begin{cases} S_2(n,m|x), & \text{if } n \geq m, \\ 0, & \text{if } n < m. \end{cases}$$
$$(35)$$

We can see this, for example, by taking $\lambda \to 0$ in (51).

Remark 1. *Combing (33) and (35), we obtain*

$$S_n(N|x) = \sum_{m=0}^{min\{N,n\}} \binom{N+1}{m+1} m! S_2(n,m|x).$$

For any $m, k \in \mathbb{N}$ with $m - k \geq 2$, we observe that

$$
\begin{aligned}
\int_{\mathbb{Z}_p} x^{m-k} d\mu_0(x) &= \int_{\mathbb{Z}_p} (x+1)^{m-k} d\mu_0(x) \\
&= \sum_{j=0}^{m-k} \binom{m-k}{m-k-j} \int_{\mathbb{Z}_p} x^{m-k-j} d\mu_0(x) \\
&= \sum_{j=k}^{m} \binom{m-k}{m-j} \int_{\mathbb{Z}_p} x^{m-j} d\mu_0(x) \\
&= \frac{1}{\binom{m}{k}} \sum_{j=k}^{m} \binom{m}{j} \binom{j}{k} \int_{\mathbb{Z}_p} x^{m-j} d\mu_0(x).
\end{aligned}
\tag{36}
$$

Thus we have shown the following result.

Theorem 4. *For any $m, k \in \mathbb{N}$ with $m - k \geq 2$, the following holds true:*

$$
\binom{m}{k} \int_{\mathbb{Z}_p} x^{m-k} d\mu_0(x) = \sum_{j=k}^{m} \binom{m}{j} \binom{j}{k} \int_{\mathbb{Z}_p} x^{m-j} d\mu_0(x).
\tag{37}
$$

From (10) and (37), we derive the following corollary.

Corollary 2. *For $m, k \in \mathbb{N}$ with $m - k \geq 2$, we have*

$$
\binom{m}{k} B_{m-k} = \sum_{j=k}^{m} \binom{m}{j} \binom{j}{k} B_{m-j}.
\tag{38}
$$

3. Some Identities of Degenerate Bernoulli Numbers and Polynomials

In this section, we assume that $0 \neq \lambda \in \mathbb{C}_p$ with $|\lambda|_p < p^{-\frac{1}{p-1}}$. The degenerate exponential function is defined as (see [3,13])

$$
e_\lambda^x(t) = (1 + \lambda t)^{\frac{x}{\lambda}}.
$$

Note that $\lim_{\lambda \to 0} e_\lambda^x(t) = e^{xt}$. In addition, we denote $(1 + \lambda t)^{\frac{1}{\lambda}} = e_\lambda^1(t)$ simply by $e_\lambda(t)$. As is well known, the degenerate Bernoulli polynomials are defined by Carlitz as

$$
\frac{t}{e_\lambda(t) - 1} e_\lambda^x(t) = \frac{t}{(1 + \lambda t)^{\frac{1}{\lambda}} - 1} (1 + \lambda t)^{\frac{x}{\lambda}} = \sum_{n=0}^{\infty} \beta_{n,\lambda}(x) \frac{t^n}{n!}.
\tag{39}
$$

When $x = 0$, $\beta_{n,\lambda} = \beta_{n,\lambda}(0)$ are called the degenerate Bernoulli numbers, (see [3,15]). From (39), we note that (see [3])

$$
\beta_{n,\lambda}(x) = \sum_{l=0}^{n} \binom{n}{l} (x)_{n-l,\lambda} \beta_{l,\lambda},
\tag{40}
$$

where $(x)_{0,\lambda} = 1$, $(x)_{n,\lambda} = x(x - \lambda) \cdots (x - (n-1)\lambda)$, $(n \geq 1)$.
By (39) and (40), we get

$$
\beta_{n,\lambda}(1) - \beta_{n,\lambda} = \delta_{n,1}.
\tag{41}
$$

Now, we observe that

$$
\begin{aligned}
\sum_{k=0}^{N} e_\lambda^{k+x}(t) &= \frac{e_\lambda^{N+1}(t)-1}{e_\lambda(t)-1} e_\lambda^x(t) = \frac{1}{t}\left\{ \frac{t}{e_\lambda(t)-1}\left(e_\lambda^{N+1+x}(t) - e_\lambda^x(t) \right) \right\} \\
&= \frac{1}{t}\sum_{n=0}^{\infty}\left(\beta_{n,\lambda}(N+1+x) - \beta_{n,\lambda}(x) \right)\frac{t^n}{n!} \\
&= \sum_{n=0}^{\infty}\left(\frac{\beta_{n+1,\lambda}(N+1+x) - \beta_{n+1,\lambda}(x)}{n+1} \right)\frac{t^n}{n!}, \quad (n \in \mathbb{N}_0).
\end{aligned}
\tag{42}
$$

On the other hand,

$$
\sum_{k=0}^{N} e_\lambda^{k+x}(t) = \sum_{n=0}^{\infty}\left(\sum_{k=0}^{N}(k+x)_{n,\lambda} \right)\frac{t^n}{n!}.
\tag{43}
$$

Let us define a degenerate version of the integer power sum polynomials, called the *degenerate integer power sum polynomials,* by

$$
S_{p,\lambda}(n|x) = \sum_{k=0}^{n}(k+x)_{p,\lambda}, \quad (n \geq 0).
\tag{44}
$$

Note that $\lim_{\lambda \to 0} S_{p,\lambda}(n|x) = S_p(n|x), \ (n \geq 0)$.

Therefore, by (42) and (43), we obtain the following theorem.

Theorem 5. *For $n, N \in \mathbb{N}_0$, we have*

$$
S_{n,\lambda}(N|x) = \frac{1}{n+1}\left(\beta_{n+1,\lambda}(N+1+x) - \beta_{n+1,\lambda}(x) \right).
\tag{45}
$$

Now, we observe that

$$
\begin{aligned}
\sum_{k=0}^{N} e_\lambda^{x+k}(t) &= \frac{1}{e_\lambda(t)-1}\left(e_\lambda^{N+1}(t) - 1 \right)e_\lambda^x(t) \\
&= \frac{1}{e_\lambda(t)-1}\left((e_\lambda(t)-1+1)^{N+1} - 1 \right)e_\lambda^x(t) \\
&= \frac{1}{e_\lambda(t)-1}\sum_{m=1}^{N+1}\binom{N+1}{m}(e_\lambda(t)-1)^m e_\lambda^x(t) \\
&= \sum_{m=0}^{N}\binom{N+1}{m+1}(e_\lambda(t)-1)^m e_\lambda^x(t) \\
&= \sum_{n=0}^{\infty}\left(\sum_{m=0}^{N}\binom{N+1}{m+1}\sum_{k=0}^{m}\binom{m}{k}(-1)^{m-k}(k+x)_{n,\lambda} \right)\frac{t^n}{n!} \\
&= \sum_{n=0}^{\infty}\left(\sum_{k=0}^{N}\sum_{m=k}^{N}\binom{N+1}{m+1}\binom{m}{k}(-1)^{m-k}(k+x)_{n,\lambda} \right)\frac{t^n}{n!}.
\end{aligned}
\tag{46}
$$

Therefore, (31) and (46) together give the next result.

Theorem 6. *For any $n, N \in \mathbb{N}_0$, the following identity holds:*

$$
S_{n,\lambda}(N|x) = \sum_{m=0}^{N}\binom{N+1}{m+1}\Delta^m (x)_{n,\lambda} = \sum_{k=0}^{N}(k+x)_{n,\lambda}T(N,k),
\tag{47}
$$

where $T(N,k) = \sum_{m=k}^{N} \binom{N+1}{m+1}\binom{m}{k}(-1)^{m-k}$.

As is known, the degenerate Stirling polynomials of the second kind are defined by Kim as (see [14])

$$(x+y)_{n,\lambda} = \sum_{k=0}^{n} S_{2,\lambda}(n,k|x)(y)_k, \tag{48}$$

where $(x)_0 = 1$, $(x)_n = x(x-1)\cdots(x-n+1)$, $(n \geq 1)$.

From (48), we can derive the generating function for $S_{2,\lambda}(n,k|x)$, $(n,k \geq 0)$, as follows:

$$\frac{1}{k!}(e_\lambda(t)-1)^k e_\lambda^x(t) = \sum_{n=k}^{\infty} S_{2,\lambda}(n,k|x)\frac{t^n}{n!}. \tag{49}$$

When $x = 0$, $S_{2,\lambda}(n,k|0) = S_{2,\lambda}(n,k)$ are called the degenerate Stirling numbers of the second kind.

By (49), we get

$$\begin{aligned}
\sum_{n=m}^{\infty} S_{2,\lambda}(n,m|x)\frac{t^n}{n!} &= \frac{1}{m!}(e_\lambda(t)-1)^m e_\lambda^x(t) \\
&= \frac{1}{m!}\sum_{k=0}^{m}\binom{m}{k}(-1)^{m-k}e_\lambda^{k+x}(t) \\
&= \sum_{n=0}^{\infty}\left(\frac{1}{m!}\sum_{k=0}^{m}\binom{m}{k}(-1)^{m-k}(x+k)_{n,\lambda}\right)\frac{t^n}{n!} \\
&= \sum_{n=0}^{\infty}\left(\frac{1}{m!}\Delta^m(x)_{n,\lambda}\right)\frac{t^n}{n!}.
\end{aligned} \tag{50}$$

Now, comparison of the coefficients on both sides of (50) yield following theorem.

Theorem 7. *For any $n,m \geq 0$, the following identity holds:*

$$\frac{1}{m!}\Delta^m(x)_{n,\lambda} = \begin{cases} S_{2,\lambda}(n,m|x), & if\ n \geq m, \\ 0, & if\ n < m. \end{cases} \tag{51}$$

Remark 2. *Combing (47) and (51), we obtain*

$$S_{n,\lambda}(N|x) = \sum_{m=0}^{\min\{N,n\}}\binom{N+1}{m+1}m!S_{2,\lambda}(n,m|x).$$

From (30) and proceeding by induction, we have

$$(1+\Delta)^k f(x) = \sum_{m=0}^{k}\binom{k}{m}\Delta^m f(x) = f(x+k), \quad (k \geq 0). \tag{52}$$

By (52), we get

$$\sum_{k=0}^{N}(x+k)_{n,\lambda} = \sum_{k=0}^{N}(1+\Delta)^k(x)_{n,\lambda}. \tag{53}$$

It is known that Daehee numbers are given by the generating function

$$\frac{\log(1+t)}{t} = \sum_{n=0}^{\infty} D_n\frac{t^n}{n!}, \quad (\text{see } [1,4,6]). \tag{54}$$

From (2), we have

$$\int_{\mathbb{Z}_p} e_\lambda^{x+y}(t)d\mu_0(y) = \frac{\frac{1}{\lambda}\log(1+\lambda t)}{e_\lambda(t)-1}e_\lambda^x(t)$$

$$= \frac{\log(1+\lambda t)}{\lambda t}\frac{t}{e_\lambda(t)-1}e_\lambda^x(t)$$

$$= \sum_{l=0}^{\infty} D_l \frac{\lambda^l t^l}{l!} \sum_{m=0}^{\infty} \beta_{m,\lambda}(x)\frac{t^m}{m!} \tag{55}$$

$$= \sum_{n=0}^{\infty} \left(\sum_{l=0}^{n} \binom{n}{l} \lambda^l D_l \beta_{n-l,\lambda}(x) \right) \frac{t^n}{n!}.$$

From (55), we have

$$\int_{\mathbb{Z}_p} (x+y)_{n,\lambda} d\mu_0(y) = \sum_{l=0}^{n} \binom{n}{l} \lambda^l D_l \beta_{n-l,\lambda}(x), \quad (n \geq 0).$$

4. Further Remark

A random variable X is a real-valued function defined on a sample space. We say that X is a continuous random variable if there exists a nonnegative function f, defined on $(-\infty, \infty)$, having the property that for any set B of real numbers (see [16,17])

$$P\{X \in B\} = \int_B f(x)dx. \tag{56}$$

The function f is called the probability density function of random variable X.

Let X be a uniform random variable on the interval (α, β). Then the probability density function f of X is given by

$$f(x) = \begin{cases} \frac{1}{\beta-\alpha}, & \text{if } \alpha < x < \beta, \\ 0, & \text{otherwise.} \end{cases} \tag{57}$$

Let X be a continuous random variable with the probability density function f. Then the expectation of X is defined by

$$E[X] = \int_{-\infty}^{\infty} xf(x)dx.$$

For any real-valued function $g(x)$, we have (see [16])

$$E[g(X)] = \int_{-\infty}^{\infty} g(x)f(x)dx. \tag{58}$$

Assume that $X_1, X_2, \cdots, X_k$ are independent uniform random variables on $(0,1)$. Then we have

$$
\begin{aligned}
E[e_\lambda^{x+X_1+X_2+\cdots+X_k}(t)] &= e_\lambda^x(t)E[e_\lambda^{X_1}(t)]E[e_\lambda^{X_2}(t)]\cdots E[e_\lambda^{X_k}(t)] \\
&= e_\lambda^x(t)\underbrace{\frac{\lambda}{\log(1+\lambda t)}(e_\lambda(t)-1)\times\cdots\times\frac{\lambda}{\log(1+\lambda t)}(e_\lambda(t)-1)}_{k-times}
\end{aligned}
$$

$$
\begin{aligned}
&= \left(\frac{\lambda t}{\log(1+\lambda t)}\right)^k \frac{k!}{t^k}\frac{1}{k!}(e_\lambda(t)-1)^k e_\lambda^x(t) \\
&= \frac{k!}{t^k}\sum_{l=0}^{\infty}B_l^{(l-k+1)}(1)\lambda^l\frac{t^l}{l!}\sum_{m=k}^{\infty}S_{2,\lambda}(m,k\mid x)\frac{t^m}{m!} \\
&= \frac{k!}{t^k}\sum_{n=k}^{\infty}\left(\sum_{m=k}^{n}\binom{n}{m}S_{2,\lambda}(m,k\mid x)B_{n-m}^{(n-m-k+1)}(1)\lambda^{n-m}\right)\frac{t^n}{n!},
\end{aligned}
\tag{59}
$$

where $B_n^{(\alpha)}(x)$ are the Bernoulli polynomials of order α, given by (see [4,7,8])

$$
\left(\frac{t}{e^t-1}\right)^\alpha e^{xt} = \sum_{n=0}^{\infty}B_n^{(\alpha)}(x)\frac{t^n}{n!},
\tag{60}
$$

and we used the well-known formula

$$
\left(\frac{t}{\log(1+t)}\right)^n (1+t)^{x-1} = \sum_{k=0}^{\infty}B_k^{(k-n+1)}(x)\frac{t^k}{k!}.
\tag{61}
$$

From (59), we note that

$$
\begin{aligned}
\binom{n}{k}&E[(x+X_1+X_2+\cdots+X_k)_{n-k,\lambda}] \\
&= \sum_{m=k}^{n}\binom{n}{m}S_{2,\lambda}(m,k\mid x)B_{n-m}^{(n-m-k+1)}(1)\lambda^{n-m}.
\end{aligned}
\tag{62}
$$

5. Conclusions

It is well-known and classical that the first n positive integer power sums can be given by an expression involving some values of Bernoulli polynomials. Here we investigated some identities on Bernoulli numbers and polynomials and those on degenerate Bernoulli numbers and polynomials, which can be deduced from certain p-adic invariant integrals on $\mathbb{Z}_p$.

In particular, we introduced the integer power sum polynomials associated with integer power sums and obtained various expressions of them. Namely, they can be given in terms of Bernoulli polynomials, difference operators, and of the Stirling polynomials of the second kind. In addition, we introduced a degenerate version of the integer power sum polynomials, called the degenerate integer power sum polynomials and were able to find several representations of them. In detail, they can be represented in terms of Carlitz degenerate Bernoulli polynomials, difference operators, and of the degenerate Stirling numbers of the second kind.

In the final section, we considered an infinite family of random variables and proved that the expectations of them are expressed in terms of the degenerate Stirling polynomials of the second and some value of higher-order Bernoulli polynomials.

Most of the results in Sections 1 and 2 are reviews of known results, other than that, we demonstrated the usefulness of the p-adic invariant integrals in the study of integer power sum polynomials. However, we emphasize that the results in Sections 3 and 4 are new. In particular, we showed that the degenerate Stirling polynomials of the second kind, introduced as a degenerate version of the Stirling polynomials of the second kind, appear naturally and meaningfully in the context of calculations of an infinite family of random variables (see (62)). We also showed that they appear in an expression of the degenerate integer power sum polynomials (Remark 2) which is a degenerate version of the integer power sum polynomials (see (26)).

We have witnessed in recent years that studying various degenerate versions of some old and new polynomials, initiated by Carlitz in the classical papers [3,15], is very productive and promising (see [3,5,14,15,18,19] and references therein). Lastly, we note that this idea of considering degenerate versions of some polynomials extended even to transcendental functions like the gamma functions (see [19]).

Author Contributions: All authors contributed equally to the manuscript, and typed, read and approved the final manuscript.

Funding: This work was supported by the National Research Foundation of Korea (NRF) grant funded by the Korea government (MEST) (No. 2017R1E1A1A03070882).

Conflicts of Interest: The authors declare that they have no competing interests.

References

1. Kim, T. q-Volkenborn integration. *Russ. J. Math. Phys.* **2002**, *9*, 288–299.
2. Schikhof, W.H. *Ultrametric Calculus: An Introduction to p-Adic Analysis*; Cambridge Studies in Advanced Mathematics, 4; Cambridge University Press: Cambridge, UK, 1984.
3. Carlitz, L. Degenerate Stirling, Bernoulli and Eulerian numbers. *Utilitas Math. J.* **1979**, *15*, 51–88.
4. El-Desoulky, B.S.; Mustafa, A. New results on higher-order Daehee and Bernoulli numbers and polynomials. *Adv. Differ. Equ.* **2016**, *2016*, 32. [CrossRef]
5. Kim, T.; Kim, D.S. Identities for degenerate Bernoulli polynomials and Korobov polynomials of the first kind. *Sci. China Math.* **2019**, *62*, 999–1028. [CrossRef]
6. Araci, S.; Özer, O. Extened q-Dedekind-type Daehee-Changhee sums associated with q-Euler polynomials. *Adv. Differ. Equ.* **2015**, *2015*, 272. [CrossRef]
7. Kim, T. Symmetry of power sum polynomials and multivariate fermionic p-adic invariant integral on $\mathbb{Z}_p$. *Russ. J. Math. Phys.* **2009**, *16*, 93–96. [CrossRef]
8. Kim, T. Sums of powers of consecutive q-integers. *Adv. Stud. Contemp. Math. (Kyungshang)* **2004**, *9*, 15–18.
9. Kim, T. A note on exploring the sums of powers of consecutive q-integers. *Adv. Stud. Contemp. Math. (Kyungshang)* **2005**, *11*, 137–140.
10. Kim, T. On the alternating sums of powers of consecutive integers. *J. Anal. Comput.* **2005**, *1*, 117–120.
11. Rim, S.-H.; Kim, T.; Ryoo, C.S. On the alternating sums of powers of consecutive q-integers. *Bull. Korean Math. Soc.* **2006**, *43*, 611–617. [CrossRef]
12. Ryoo, C.S.; Kim, T. Exploring the q-analogues of the sums of powers of consecutive integers with Mathematica. *Adv. Stud. Contemp. Math. (Kyungshang)* **2009**, *18*, 69–77.
13. Simsek, Y.; Kim, D.S.; Kim, T.; Rim, S.-H. A note on the sums of powers of consecutive q-integers. *J. Appl. Funct. Differ. Equ.* **2016**, *1*, 81–88.
14. Kim, T. A note on degenerate Stirling polynomials of the second kind. *Proc. Jangjeon Math. Soc.* **2017**, *20*, 319–331.
15. Carlitz, L. A degenerate Staudt-Clausen theorem. *Arch. Math. (Basel)* **1956**, *7*, 28–33. [CrossRef]
16. Kim, T.; Yao, Y.; Kim, D.S.; Kwon, H.I. Some identities involving special numbers and moments of random variables. *Rocky Mt. J. Math.* **2019**, *49*, 521–538. [CrossRef]
17. Liu, C.; Bao, W. Application of probabilistic method on Daehee sequences. *Eur. J. Pure Appl. Math.* **2018**, *11*, 69–78. [CrossRef]

Symmetry **2019**, *11*, 847

18. Kim, T.; Yao, Y.; Kim, D.S.; Jang, G.-W. Degenerate *r*-Stirling numbers and *r*-Bell polynomials. *Russ. J. Math. Phys.* **2018**, *25*, 44–58. [CrossRef]
19. Kim, T.; Kim, D.S. Degenerate Laplace transform and degenerate gamma function. *Russ. J. Math. Phys.* **2017**, *24*, 241–248. [CrossRef]

 *symmetry*

Article

Some Convolution Formulae Related to the Second-Order Linear Recurrence Sequence

Zhuoyu Chen [1] and Lan Qi [2,*]

[1]   School of Mathematics, Northwest University, Xi'an 710127, China; chenzymath@stumail.nwu.edu.cn
[2]   School of Mathematics and Statistics, Yulin University, Yulin 719000, China
*   Correspondence: qilanmail@163.com

Received: 9 May 2019; Accepted: 12 June 2019; Published: 13 June 2019

Abstract: The main aim of this paper is that for any second-order linear recurrence sequence, the generating function of which is $f(t) = \frac{1}{1+at+bt^2}$, we can give the exact coefficient expression of the power series expansion of $f^x(t)$ for $x \in \mathbf{R}$ with elementary methods and symmetry properties. On the other hand, if we take some special values for a and b, not only can we obtain the convolution formula of some important polynomials, but also we can establish the relationship between polynomials and themselves. For example, we can find relationship between the Chebyshev polynomials and Legendre polynomials.

Keywords: Fibonacci numbers; Lucas numbers; Chebyshev polynomials; Legendre polynomials; Jacobi polynomials; Gegenbauer polynomials; convolution formula

MSC: 11B83

---

1. Introduction

For any integer $n \geq 1$ and any real number y, the Fibonacci polynomials $F_n(y)$ and the Lucas polynomials $L_n(y)$ are defined by the second-order linear recurrence sequence

$$F_{n+1}(y) = yF_n(y) + F_{n-1}(y)$$

and

$$L_{n+1}(y) = yL_n(y) + L_{n-1}(y),$$

where the first two terms are $F_0(y) = 0$, $F_1(y) = 1$, $L_0(y) = 2$ and $L_1(y) = y$.

If we take $\alpha = \frac{y+\sqrt{y^2+4}}{2}$, $\beta = \frac{y-\sqrt{y^2+4}}{2}$, according to the properties of the second-order linear recurrence sequence, we have

$$F_n(y) = \frac{\alpha^n - \beta^n}{\alpha - \beta}$$

and

$$L_n(y) = \alpha^n + \beta^n.$$

For any integer $n \geq 0$, the Fibonacci numbers $F_n = F_n(1)$ can be defined by the generating function

$$\frac{1}{1-t-t^2} = \sum_{n=0}^{\infty} F_n t^n.$$

For any integer $n \geq 0$, the first and the second kind Chebyshev polynomials $T_n(y)$ and $U_n(y)$ are defined by the second-order linear recurrence sequence

$$T_{n+2}(y) = 2yT_{n+1}(y) - T_n(y)$$

and

$$U_{n+2}(y) = 2yU_{n+1}(y) - U_n(y),$$

where the first two terms are $T_0(y) = 1$, $T_1(y) = y$, $U_0(y) = 1$ and $U_1(y) = 2y$.

If we take $\alpha = y + \sqrt{y^2 - 1}$, $\beta = y - \sqrt{y^2 - 1}$, according to the properties of the second-order linear recurrence sequence, we have

$$T_n(y) = \frac{\alpha^n + \beta^n}{2}$$

and

$$U_n(y) = \frac{\alpha^{n+1} - \beta^{n+1}}{\alpha - \beta}.$$

On the other hand, the second kind Chebyshev polynomials $U_n(y)$ can be also defined by the generating function

$$\frac{1}{1 - 2yt + t^2} = \sum_{n=0}^{\infty} U_n(y)t^n.$$

Besides Fibonacci polynomials, Lucas polynomials and Chebyshev polynomials, other orthogonal polynomials have also been studied by interested scholars.

For example, the Legendre polynomials $P_n(y)$ are defined by the generating function

$$\left(\frac{1}{1 - 2yt + t^2}\right)^{\frac{1}{2}} = \sum_{n=0}^{\infty} P_n(y)t^n.$$

The Jacobi polynomials $\{P_n^{(\alpha,\beta)}(y)\}_{0 \leq n < \infty}$ are defined by the generating function

$$\left[R(1 + R - t)^{\alpha}(1 + R + t)^{\beta}\right]^{-1} = \sum_{k=0}^{\infty} 2^{-\alpha-\beta} P_n^{(\alpha,\beta)}(y)t^n,$$

where $R = \sqrt{1 - 2yt + t^2}$, $|t| < 1$, $\alpha, \beta > -1$.

The Gegenbauer polynomials $\{C_n^{\lambda}(y)\}_{0 \leq n < \infty}$ are defined by the generating function

$$\left(\frac{1}{1 - 2yt + t^2}\right)^{\lambda} = \sum_{n=0}^{\infty} C_n^{\lambda}(y)t^n, \left(\lambda > -\frac{1}{2}\right).$$

It is well know that polynomials and sequence occupy indispensable positions in the research of number theory. Especially, Fibonacci and Lucas numbers, Chebyshev and Legendre polynomials and others. These polynomials and numbers are closely related and there are a variety of meaningful results which have been researched by interested scholars until now. For example, the identities of Chebyshev polynomials can be found in [1–9], and the contents about Fibonacci and Lucas numbers in [10,11]. Some authors have a research which connects Chebyshev polynomials and Fibonacci or Lucas polynomials (see [12–14]).

In particular, we can find many significant results in the aspect of studying the calculating problem of one kind sums of some important polynomials. For example, Yuankui Ma and Wenpeng Zhang have calculated one kind sums of Fibonacci Polynomials (see [15]) as follows.

Let h be a positive integer, for any integer $n \geq 0$, they proved

$$\sum_{a_1+a_2+\cdots+a_{h+1}=n} F_{a_1}(x)F_{a_2}(x)\cdots F_{a_{h+1}}(x) = \frac{1}{h!}\cdot\sum_{j=1}^{h}\frac{(-1)^{h-j}\cdot S(h,j)}{x^{2h-j}}$$
$$\times\left(\sum_{i=0}^{n}\frac{(n-i+j)!}{(n-i)!}\cdot\binom{2h+i-j-1}{i}\cdot\frac{(-1)^i\cdot 2^i\cdot F_{n-i+j}(x)}{x^i}\right),$$

where the summation is over all $h+1$-tuples with non-negative integer coordinates $(a_1, a_2, \cdots, a_{h+1})$ such that $a_1 + a_2 + \cdots + a_{h+1} = n$, and $S(h,i)$ is a second order non-linear recurrence sequence defined by $S(h,0) = 0$, $S(h,h) = 1$, and $S(h+1, i+1) = 2\cdot(2h-1-i)\cdot S(h, i+1) + S(h,i)$ for all positive integers $1 \leq i \leq h-1$.

Yixue Zhang and Zhuoyu Chen have researched the calculating problem of one kind sums of the second kind Chebyshev polynomials (see [16]) as follows.

Let h be a positive integer, for any integer $n \geq 0$, they proved

$$\sum_{a_1+a_2+\cdots+a_{h+1}=n} U_{a_1}(x)U_{a_2}(x)\cdots U_{a_{h+1}}(x)$$
$$= \frac{1}{2^h\cdot h!}\cdot\sum_{j=1}^{h}\frac{C(h,j)}{x^{2h-j}}\sum_{i=0}^{n}\frac{(n-i+j)!}{(n-i)!}\cdot\binom{2h+i-j-1}{i}\cdot\frac{U_{n-i+j}(x)}{x^i},$$

where $C(h,i)$ is a second order non-linear recurrence sequence defined by $C(h,0) = 0$, $C(h,h) = 1$, $C(h+1,1) = 1\cdot 3\cdot 5\cdots(2h-1) = (2h-1)!!$ and $C(h+1, i+1) = (2h-1-i)\cdot C(h, i+1) + C(h,i)$ for all $1 \leq i \leq h-1$.

Shimeng Shen and Li Chen have studied the calculating problem of one kind sums of Legendre Polynomials (see [17]) as follows.

For any positive integer k and integer $n \geq 0$, they proved

$$(2k-1)!!\sum_{a_1+a_2+\cdots+a_{2k+1}=n} P_{a_1}(x)P_{a_2}(x)\cdots P_{a_k}(x)$$
$$= \sum_{j=1}^{k}C(k,j)\sum_{i=0}^{n}\frac{(n+k+1-i-j)!}{(n-i)!}\cdot\frac{\binom{i+j+k-2}{i}}{x^{k-1+i+j}}\cdot P_{n+k+1-i-j}(x)$$

where $(2k-1)!! = 1\times 3\times 5\cdots(2k-1) = 2^k(\frac{1}{2})_k$, and $C(k,i)$ is a recurrence sequence defined by $C(k,1) = 1$, $C(k+1,k+1) = (2k-1)!!$ and $C(k+1, i+1) = C(k, i+1) + (k-1+i)\cdot C(k,i)$ for all $1 \leq i \leq k-1$.

They have converted the complex sums of $F_n(x)$ into a simple combination of $F_n(x)$, the complex sums of $U_n(x)$ into a simple combination of $U_n(x)$, and the complex sums of $P_n(x)$ into a simple combination of $P_n(x)$.

Very recently, Taekyun Kim and other people researched the properties of Fibonacci numbers through introducing the convolved Fibonacci numbers $p_n(x)$ by generating function as follows (see [18]):

$$\left(\frac{1}{1-t-t^2}\right)^x = \sum_{n=0}^{\infty} p_n(x)\frac{t^n}{n!}, (x\in\mathbf{R}).$$

They researched some new and explicit identities of the convolved Fibonacci numbers for $x\in\mathbf{N}$. For example, for $n\geq 0$ and $r\in\mathbf{N}$, they have proved the recurrence relationship of $p_n(x)$ (see [18]):

$$p_n(x) = \sum_{l=0}^{n} p_l(r)p_{n-l}(x-r) = \sum_{l=0}^{n} p_{n-l}(r)p_l(x-r).$$

The convolved Fibonacci numbers $p_n(x)$ seems to be only connected with the simple power square. In fact, it can establish the relationship between polynomials and themselves, so the further research of $p_n(x)$ is very significant. They have provided us a new perspective to study the properties of some vital polynomials. For example, Taekyun Kim and other people have proved the relationship between $p_n(x)$ and the combination sums about Fibonacci numbers:

$$\frac{p_n(r+1)}{n!} = \sum_{l_1=0}^{n} \sum_{l_2=0}^{n-l_1} \cdots \sum_{l_r=0}^{n-l_1-\cdots-l_{r-1}} F_{l_1} F_{l_2} \cdots F_{l_r} F_{n-l_1-l_2-\cdots-l_r}.$$

They have converted the complex sums of $F_n(x)$ into a calculation problem of $p_n(x)$ and the calculation method is easier and the expression is simpler.

Inspired by this article, in this paper, for any second-order linear recurrence sequence, the generating function of which is $f(t) = \frac{1}{1+at+bt^2}$, we can define

$$\left(\frac{1}{1+at+bt^2}\right)^x = \sum_{n=0}^{\infty} p_n(x)\frac{t^n}{n!}, (a, b, x \in \mathbf{R}). \tag{1}$$

Firstly, we give a specific computational formula of $p_n(x)$ for $x \in \mathbf{R}$ using the elementary methods. After that for any polynomial or sequence, the generating function of which is $f(t) = \frac{1}{1+at+bt^2}$, we can obtain its convolved formula easily and directly.

Secondly, if we take some special values for a, b in $f(t)$ and x in $p_n(x)$, we can find some relationship between special polynomials and themselves. For example, we will establish the relationship between the convolved Fibonacci numbers and Lucas numbers, the relationship between the convolved formula of the second kind Chebyshev polynomials and the first kind Chebyshev polynomials, and the relationship between Legendre polynomials and the first kind Chebyshev polynomials and others.

At last, through the computational formula of $p_n(x)$, especially for $x \in \mathbf{N}$, we can also convert the complex sums of F_n into a liner combination of L_n; and express the complex sums of $U_n(y)$ as a liner combination of $T_n(y)$. More importantly, the forms are more common and the calculations are easier than previous results.

We will prove the main results as follows:

Theorem 1. *Let* $f(t) = \frac{1}{1-t-t^2}$, *for any integer* $n \geq 0$ *and* $x \in \mathbf{R}$, *we can obtain*

$$p_n(x) = \frac{1}{2} \sum_{i=0}^{n} (-1)^i \binom{n}{i} \langle x \rangle_i \langle x \rangle_{n-i} L_{n-2i},$$

where $\langle x \rangle_n = x(x+1)(x+2)\cdots(x+n-1)$ *and* $\langle x \rangle_0 = 1$.

Theorem 2. *Let* $f(t) = \frac{1}{1-2yt+t^2}$, *for any integer* $n \geq 0$ *and* $x, y \in \mathbf{R}$, *we can obtain*

$$p_n(x;y) = \sum_{i=0}^{n} \binom{n}{i} \langle x \rangle_i \langle x \rangle_{n-i} T_{n-2i}(y).$$

From Theorem 1 we can deduce the following:

Corollary 1. *For any positive integer k, we have the identity*

$$\sum_{a_1+a_2+\cdots+a_k=n} F_{a_1} F_{a_2} \cdots F_{a_k}$$

$$= \frac{1}{2((k-1)!)^2} \sum_{i=0}^{n} (-1)^i \frac{(k+i-1)!(k+n-i-1)!}{i!(n-i)!} \cdot L_{n-2i}.$$

From Theorem 2 we can deduce the following:

Corollary 2. *For any positive integer k, we have the identity*

$$\sum_{a_1+a_2+\cdots+a_k=n} U_{a_1}(y) \cdot U_{a_2}(y) \cdots U_{a_k}(y)$$

$$= \frac{1}{((k-1)!)^2} \sum_{i=0}^{n} \frac{(k+i-1)!(k+n-i-1)!}{i!(n-i)!} \cdot T_{n-2i}(y).$$

Corollary 3. *If $x = \frac{1}{2}$, we have the identity*

$$P_n(y) = \frac{1}{2^n} \sum_{i=0}^{n} \frac{(2i-1)!!(2n-2i-1)!!}{i!(n-i)!} \cdot T_{n-2i}(y).$$

Corollary 4. *If $x = -\frac{1}{2}$, we have the identity*

$$R = \sum_{n=0}^{\infty} \frac{1}{2^n} \sum_{i=0}^{n} \frac{(2i-3)!!(2n-2i-3)!!}{i!(n-i)!} \cdot T_{n-2i}(y) \cdot t^n.$$

Corollary 5. *If $x = \lambda > -\frac{1}{2}$, we have the identity*

$$C_n^{\lambda}(y) = \frac{1}{n!} \sum_{i=0}^{n} \binom{n}{i} \langle \lambda \rangle_i \langle \lambda \rangle_{n-i} T_{n-2i}(y).$$

Theorems 1 and 2 give the computational formula of $p_n(x)$ of some famous polynomials. Especially, we know that polynomials are closely connected and they can be converted to each other. According to these theorems, we can obtain the relationship between the polynomials easily. It cannot only extend the application of orthogonal polynomials, but also make replacement calculations according to its complexity. For example, if we make a calculation involving the Gegenbauer polynomials, for simple calculations, we can convert it into Chebyshev polynomials according to Corollary 5.

2. A Simple Lemma

In order to prove our theorems, we are going to introduce a simple lemma.

Lemma 1. *For any integer $n \geq 0$ and $a, b, x \in \mathbf{R}$, we can obtain the equation*

$$p_n(x) = \frac{1}{2} \sum_{i=0}^{n} b^i \binom{n}{i} \langle x \rangle_i \langle x \rangle_{n-i} \left(\left(\frac{-a + \sqrt{a^2 - 4b}}{2} \right)^{n-2i} + \left(\frac{-a - \sqrt{a^2 - 4b}}{2} \right)^{n-2i} \right).$$

Proof. Firstly, according Equation (1), we have

$$\sum_{n=0}^{\infty} p_n(x) \frac{t^n}{n!} = \left(\frac{1}{1 + at + bt^2} \right)^x = (1 - \alpha t)^{-x} (1 - \beta t)^{-x}. \tag{2}$$

We can easily know that $\alpha + \beta = -a$, $\alpha\beta = b$ and $\alpha = \frac{-a+\sqrt{a^2-4b}}{2}$, $\beta = \frac{-a-\sqrt{a^2-4b}}{2}$ are two roots of $1 + at + bt^2 = 0$.

Then, applying the properties of power series, we obtain

$$(1-\alpha t)^{-x} = \sum_{n=0}^{\infty} \binom{-x}{n}(-1)^n(\alpha t)^n = \sum_{n=0}^{\infty} \frac{(-x)_n}{n!}(-1)^n \alpha^n t^n \tag{3}$$

and

$$(1-\beta t)^{-x} = \sum_{n=0}^{\infty} \binom{-x}{n}(-1)^n(\beta t)^n = \sum_{n=0}^{\infty} \frac{(-x)_n}{n!}(-1)^n \beta^n t^n, \tag{4}$$

where $(x)_n = x(x-1)(x-2)\cdots(x-n+1)$ and $(x)_0 = 1$.

Combining Equations (2)–(4), we get

$$
\begin{aligned}
\sum_{n=0}^{\infty} p_n(x)\frac{t^n}{n!} &= \left(\sum_{n=0}^{\infty} \frac{(-x)_n}{n!}(-1)^n \alpha^n t^n\right)\left(\sum_{n=0}^{\infty} \frac{(-x)_n}{n!}(-1)^n \beta^n t^n\right)\\
&= \sum_{n=0}^{\infty}\left(\sum_{i=0}^{n} \frac{(-x)_i(-1)^i\alpha^i t^i}{i!}\cdot\frac{(-x)_{n-i}(-1)^{n-i}\beta^{n-i}t^{n-i}}{(n-i)!}\right)\\
&= \sum_{n=0}^{\infty} \frac{(-1)^n}{n!}\left(\sum_{i=0}^{n}\binom{n}{i}(-x)_i(-x)_{n-i}\alpha^i\beta^{n-i}\right)t^n.
\end{aligned}\tag{5}
$$

Similarly, according the symmetry of α and β, we can easily obtain

$$\sum_{n=0}^{\infty} p_n(x)\frac{t^n}{n!} = \sum_{n=0}^{\infty} \frac{(-1)^n}{n!}\left(\sum_{i=0}^{n}\binom{n}{i}(-x)_i(-x)_{n-i}\beta^i\alpha^{n-i}\right)t^n. \tag{6}$$

Then, combining Equations (5) and (6), we know that

$$
\begin{aligned}
\sum_{n=0}^{\infty} p_n(x)\frac{t^n}{n!} &= \frac{1}{2}\sum_{n=0}^{\infty} \frac{(-1)^n}{n!}\left(\sum_{i=0}^{n}\binom{n}{i}(-x)_i(-x)_{n-i}(\alpha\beta)^i\left(\beta^{n-2i}+\alpha^{n-2i}\right)\right)t^n\\
&= \frac{1}{2}\sum_{n=0}^{\infty} \frac{1}{n!}\left(\sum_{i=0}^{n} b^i\binom{n}{i}\langle x\rangle_i\langle x\rangle_{n-i}\left(\beta^{n-2i}+\alpha^{n-2i}\right)\right)t^n.
\end{aligned}\tag{7}
$$

Comparing the coefficients of t^n in Equation (7), we get

$$
\begin{aligned}
p_n(x) &= \frac{1}{2}\sum_{i=0}^{n} b^i\binom{n}{i}\langle x\rangle_i\langle x\rangle_{n-i}\left(\alpha^{n-2i}+\beta^{n-2i}\right)\\
&= \frac{1}{2}\sum_{i=0}^{n} b^i\binom{n}{i}\langle x\rangle_i\langle x\rangle_{n-i}\left(\left(\frac{-a+\sqrt{a^2-4b}}{2}\right)^{n-2i}+\left(\frac{-a-\sqrt{a^2-4b}}{2}\right)^{n-2i}\right).
\end{aligned}
$$

Now we have completed the proof of the Lemma 1. $\square$

3. Proof of the Theorem

Proof of Theorem 1. If we take $a = -1$ and $b = -1$ in Equation (1), we know that $f(t)$ is the generating function of Fibonacci number. That is,

$$f(t) = \frac{1}{1-t-t^2} = \sum_{n=0}^{\infty} F_n t^n.$$

The convolved Fibonacci numbers $p_n(x)$ are defined by the generating function as [18]

$$f^x(t) = \left(\frac{1}{1-t-t^2}\right)^x = \sum_{n=0}^{\infty} p_n(x)\frac{t^n}{n!}. \tag{8}$$

In this time, $\alpha = \frac{1+\sqrt{5}}{2}, \beta = \frac{1-\sqrt{5}}{2}$.
According to the Lemma 1 and $L_n = \alpha^n + \beta^n$, we can get

$$
\begin{aligned}
p_n(x) &= \frac{1}{2}\sum_{i=0}^{n} b^i \binom{n}{i} \langle x\rangle_i \langle x\rangle_{n-i} \left(\left(\frac{-a+\sqrt{a^2-4b}}{2}\right)^{n-2i} + \left(\frac{-a-\sqrt{a^2-4b}}{2}\right)^{n-2i}\right) \\
&= \frac{1}{2}\sum_{i=0}^{n} b^i \binom{n}{i} \langle x\rangle_i \langle x\rangle_{n-i} \left(\left(\frac{1+\sqrt{5}}{2}\right)^{n-2i} + \left(\frac{1-\sqrt{5}}{2}\right)^{n-2i}\right) \\
&= \frac{1}{2}\sum_{i=0}^{n} b^i \binom{n}{i} \langle x\rangle_i \langle x\rangle_{n-i} \left(\alpha^{n-2i} + \beta^{n-2i}\right) \\
&= \frac{1}{2}\sum_{i=0}^{n} (-1)^i \binom{n}{i} \langle x\rangle_i \langle x\rangle_{n-i} L_{n-2i}.
\end{aligned} \tag{9}
$$

In this equation, $p_n(x)$ is expressed as a combined forms of Lucas number. The Proof of Theorem 1 has finished. $\square$

About the convolved Fibonacci numbers $p_n(x)$, Taekyun Kim and others have obtained its some-recurrence formulae in reference [18]. Based on [18], we have given an exact computational formula of $p_n(x)$ for any arbitrary x in Theorem 1. Compared with the results in [18], Theorem 1 is more general and easier.

If we take $x = k \in \mathbf{N}$ in Equation (8), we get

$$
\begin{aligned}
\sum_{n=0}^{\infty} p_n(k)\frac{t^n}{n!} &= \left(\frac{1}{1-t-t^2}\right)^k = \left(\sum_{n=0}^{\infty} F_n t^n\right)^k \\
&= \left(\sum_{a_1=0}^{\infty} F_{a_1}\cdot t^{a_1}\right)\left(\sum_{a_2=0}^{\infty} F_{a_2}\cdot t^{a_2}\right)\cdots\left(\sum_{a_k=0}^{\infty} F_{a_k}\cdot t^{a_k}\right) \\
&= \left(\sum_{a_1=0}^{\infty}\sum_{a_2=0}^{\infty}\cdots\sum_{a_k=0}^{\infty} F_{a_1}\cdot F_{a_2}\cdots F_{a_k}\cdot t^{a_1+a_2\cdots+a_k}\right) \\
&= \sum_{n=0}^{\infty}\left(\sum_{a_1+a_2+\cdots+a_k=n} F_{a_1}\cdot F_{a_2}\cdots F_{a_k}\right)\cdot t^n,
\end{aligned}
$$

and then combining Equation (9), we can obtain

$$
\begin{aligned}
\sum_{a_1+a_2+\cdots+a_k=n} & F_{a_1}\cdot F_{a_2}\cdots F_{a_k} \\
&= \frac{1}{2n!}\sum_{i=0}^{n} (-1)^i \binom{n}{i} \langle k\rangle_i \langle k\rangle_{n-i} L_{n-2i} \\
&= \frac{1}{2((k-1)!)^2}\sum_{i=0}^{n} (-1)^i \frac{(k+i-1)!(k+n-i-1)!}{i!(n-i)!} L_{n-2i}.
\end{aligned}
$$

The proof of Corollary 1 has finished.
For every $F_{a_l} (1 \le l \le k)$, $\sum_{a_1+a_2+\cdots+a_k=n} F_{a_1}\cdot F_{a_2}\cdots F_{a_k}$ is symmetry.

Proof of Theorem 2. If we take $a = -2y$ and $b = 1$ in Equation (1), we all know $f(t; y)$ is the generating function of the second-kind Chebyshev polynomials $U_n(y)$

$$f(t; y) = \frac{1}{1 - 2yt + t^2} = \sum_{n=0}^{\infty} U_n(y) t^n.$$

The convolved second-kind Chebyshev polynomials $p_n(x; y)$ are defined by the generating function as [18]

$$f^x(t; y) = \left(\frac{1}{1 - 2yt + t^2} \right)^x = \sum_{n=0}^{\infty} p_n(x; y) \frac{t^n}{n!}. \tag{10}$$

In this time, $\alpha = y + \sqrt{y^2 - 1}$, $\beta = y - \sqrt{y^2 - 1}$.

According to the Lemma 1 and $T_n(y) = \frac{1}{2}(\alpha^n + \beta^n)$, we can get

$$p_n(x; y) = \sum_{i=0}^{n} \binom{n}{i} \langle x \rangle_i \langle x \rangle_{n-i} T_{n-2i}(y). \tag{11}$$

In this equation, $p_n(x; y)$ is expressed as a combined form of the first-kind Chebyshev polynomials $T_n(x)$. $\square$

If we take $x = k \in \mathbf{N}$ in Equation (10), and combining Equation (11) we can easily prove the Corollary 2.

Take $x = \frac{1}{2}$ in Equation (10), we know $f^{\frac{1}{2}}(t; y)$ is the generating function of the Legendre polynomials $P_n(x)$ as follows:

$$f^{\frac{1}{2}}(t) = \left(\frac{1}{1 - 2yt + t^2} \right)^{\frac{1}{2}} = \sum_{n=0}^{\infty} P_n(y) t^n = \sum_{n=0}^{\infty} p_n \left(\frac{1}{2}; y \right) \frac{t^n}{n!}.$$

According to Theorem 2, we can easily obtain

$$\begin{aligned}
p_n \left(\frac{1}{2}; y \right) &= \sum_{i=0}^{n} \binom{n}{i} \left\langle \frac{1}{2} \right\rangle_i \left\langle \frac{1}{2} \right\rangle_{n-i} T_{n-2i}(y) \\
&= \frac{1}{2^{2n}} \sum_{i=0}^{n} \frac{n!(2i)!(2(n-i))!}{(i!)^2((n-i)!)^2} \cdot T_{n-2i}(y) \\
&= \frac{n!}{2^n} \sum_{i=0}^{n} \frac{(2i-1)!!(2n-2i-1)!!}{i!(n-i)!} \cdot T_{n-2i}(y).
\end{aligned}$$

In a word, we know the Legendre polynomials $P_n(x)$ can be expressed as combined forms of the first kind Chebyshev polynomials $T_n(x)$ as follows:

$$P_n(y) = \frac{1}{2^n} \sum_{i=0}^{n} \frac{(2i-1)!!(2n-2i-1)!!}{i!(n-i)!} \cdot T_{n-2i}(y).$$

The proof of Corollary 3 has finished.

If we take $x = -\frac{1}{2}$ in (10), then we can easily obtain

$$\begin{aligned}
R &= \sum_{n=0}^{\infty} p\left(-\frac{1}{2}; y \right) \frac{t^n}{n!} = \sum_{n=0}^{\infty} \frac{1}{2^{2n-2}} \sum_{i=0}^{n} \frac{(2(i-1))!(2(n-i-1))!}{i!(i-1)!(n-i)!(n-i-1)!} \cdot T_{n-2i}(y) \cdot t^n \\
&= \sum_{n=0}^{\infty} \frac{1}{2^n} \sum_{i=0}^{n} \frac{(2i-3)!!(2n-2i-3)!!}{i!(n-i)!} \cdot T_{n-2i}(y) \cdot t^n.
\end{aligned}$$

The proof of Corollary 4 has finished.

Taking $x = \lambda > -\frac{1}{2}$ in Equation (10), we know $f^\lambda(t; y)$ is the generating function of the Gegenbauer polynomials $\{C_n^\lambda(y)\}_{0 \leq n < \infty}$ as follows:

$$\left(\frac{1}{1 - 2yt + t^2}\right)^\lambda = \sum_{n=0}^{\infty} C_n^\lambda(y) t^n = f^\lambda(t; y) = \sum_{n=0}^{\infty} p_n(\lambda; y) \frac{t^n}{n!}.$$

According to Theorem 2, we can easily obtain

$$p_n(\lambda; y) = \sum_{i=0}^{n} \binom{n}{i} \langle \lambda \rangle_i \langle \lambda \rangle_{n-i} T_{n-2i}(y).$$

The proof of Corollary 5 has finished.

Author Contributions: Writing—original draft: Z.C.; Writing—review and editing: L.Q.

Funding: This work is supported by the N. S. F. (11771351), (11826205), (11826203) of P. R. China and N.S.B.R.P in Shaanxi Province (2018JQ1093).

Acknowledgments: The authors would like to thank the referees for their very helpful and detailed comments, which have significantly improved the presentation of this paper.

References

1. Chen, L.; Zhang, W. Chebyshev polynomials and their some interesting applications. *Adv. Differ. Equ.* **2017**, *2017*, 303.
2. Wang, S. Some new identities of Chebyshev polynomials and their applications. *Adv. Differ. Equ.* **2015**, *2015*, 335.
3. Li, X. Some identities involving Chebyshev polynomials. *Math. Probl. Eng.* **2015**, *2015*, 950695. [CrossRef]
4. Ma, Y.; Lv, X. Several identities involving the reciprocal sums of Chebyshev polynomials. *Math. Probl. Eng.* **2017**, *2017*, 4194579. [CrossRef]
5. Cesarano, C. Identities and generating functions on Chebyshev polynomials. *Georgian Math. J.* **2012**, *19*, 427–440. [CrossRef]
6. Cesarano, C. Integral representations and new generating functions of Chebyshev polynomials. *Hacet. J. Math. Stat.* **2015**, *44*, 535–546. [CrossRef]
7. Lee, C.; Wong, K. On Chebyshev's Polynomials and Certain Combinatorial Identities. *Bull. Malays. Math. Sci. Soc.* **2011**, *34*, 279–286.
8. Wang, T.; Zhang, H. Some identities involving the derivative of the first kind Chebyshev polynomials. *Math. Probl. Eng.* **2015**, *2015*, 146313. [CrossRef]
9. Wang, T.; Zhang, W. Two identities involving the integral of the first kind Chebyshev polynomials. *Bull. Math. Soc. Sci. Math. Roum.* **2017**, *108*, 91–98.
10. Zhang, W. Some identities involving the Fibonacci numbers and Lucas numbers. *Fibonacci Q.* **2004**, *42*, 149–154.
11. Ma, R.; Zhang, W. Several identities involving the Fibonacci numbers and Lucas numbers. *Fibonacci Q.* **2007**, *45*, 164–170.
12. Kim, T.; Kim, D.; Dolgy, D.; Park, J. Sums of finite products of Chebyshev polynomials of the second kind and of Fibonacci polynomials. *J. Inequal. Appl.* **2018**, *2018*, 148. [CrossRef] [PubMed]
13. Kim, T.; Kim, D.; Kwon, J.; Dolgy, D. Expressing Sums of Finite Products of Chebyshev Polynomials of the Second Kind and of Fibonacci Polynomials by Several Orthogonal Polynomials. *Mathematics* **2018**, *6*, 210. [CrossRef]
14. Kim, T.; Kim, D.; Dolgy, D.; Kwon, J. Representing Sums of Finite Products of Chebyshev Polynomials of the First Kind and Lucas Polynomials by Chebyshev Polynomials. *Mathematics* **2019**, *7*, 26. [CrossRef]
15. Ma, Y.; Zhang, W. Some Identities Involving Fibonacci Polynomials and Fibonacci Numbers. *Mathematics* **2018**, *6*, 334. [CrossRef]

16. Zhang, Y.; Chen, Z. A New Identity Involving the Chebyshev Polynomials. *Mathematics* **2018**, *6*, 244. [CrossRef]
17. Shen, S.; Chen, L. Some Types of Identities Involving the Legendre Polynomials. *Mathematics* **2019**, *7*, 114. [CrossRef]
18. Kim, T.; Dolgy, D.; Kim, D.; Seo, J. Convolved Fibonacci Numbers and Their Applications. *ARS Comb.* **2017**, *135*, 119–131.

Article

On r-Central Incomplete and Complete Bell Polynomials

Dae San Kim [1], Han Young Kim [2], Dojin Kim [3,*] and Taekyun Kim [2]

[1]   Department of Mathematics, Sogang University, Seoul 04107, Korea; dskim@sogang.ac.kr
[2]   Department of Mathematics, Kwangwoon University, Seoul 01897, Korea; gksdud213@kw.ac.kr (H.Y.K.);
    tkkim@kw.ac.kr (T.K.)
[3]   Department of Mathematics, Pusan National University, Busan 46241, Korea
*   Correspondence: kimdojin@pusan.ac.kr

Received: 16 April 2019; Accepted:23 May 2019; Published: 27 May 2019

Abstract: Here we would like to introduce the extended r-central incomplete and complete Bell polynomials, as multivariate versions of the recently studied extended r-central factorial numbers of the second kind and the extended r-central Bell polynomials, and also as multivariate versions of the r-Stirling numbers of the second kind and the extended r-Bell polynomials. In this paper, we study several properties, some identities and various explicit formulas about these polynomials and their connections as well.

Keywords: extended r-central complete bell polynomials; extended r-central incomplete bell polynomials; complete r-Bell polynomials; incomplete r-bell polynomials

---

1. Introduction

We begin this section by briefly recalling several definitions related to the central factorial numbers of the second kind and the central Bell polynomials and also to their generalizations of the extended r-central factorial numbers of the second kind and the extended r-central Bell polynomials (see [1]). The central factorial $x^{[n]}$ is given by the generating function

$$\left(\frac{t}{2}+\sqrt{1+\frac{t^2}{4}}\right)^{2x}=\sum_{n=0}^{\infty}x^{[n]}\frac{t^n}{n!}. \tag{1}$$

A proof of (1) can be found in [2], p. 215, Equations (27), (28) and (27a), (see also [1,3–5]).

It is well known that Formula (1) shows that

$$x^{[0]}=1,\quad x^{[n]}=x(x+\frac{n}{2}-1)\cdots(x-\frac{n}{2}+1),\quad (n\geq 1), \tag{2}$$

where $x^{[n]}$ is of degree n in x.

The central factorial numbers of the second kind $T(n,k)$ are the coefficients in the expansion of x^n in terms of central factorials as follows:

$$x^n=\sum_{k=0}^{n}T(n,k)x^{[k]}, \tag{3}$$

(see [6–11]) and it is known that $T(2n,2n-2k)$ enumerates the number of ways to place k rooks on a 3D-triangle board of size $(n-1)$ (see [12,13]). The generating function of $T(n,k)$ is given by

$$\frac{1}{k!}\left(e^{\frac{t}{2}}-e^{-\frac{t}{2}}\right)^k = \sum_{n=k}^{\infty} T(n,k)\frac{t^n}{n!},$$ (4)

which follows, for example, from (1) and (3).

Indeed, on the one hand by making use of (3) we have

$$e^{xt} = \sum_{n=0}^{\infty} x^n \frac{t^n}{n!} = \sum_{k=0}^{\infty}\left(\sum_{n=k}^{\infty} T(n,k)\frac{t^n}{n!}\right)x^{[k]}.$$ (5)

On the other hand, by virtue of (1) we also have

$$e^{xt} = \left(\frac{e^{\frac{t}{2}}-e^{-\frac{t}{2}}}{2} + \sqrt{1+\frac{\left(e^{\frac{t}{2}}-e^{-\frac{t}{2}}\right)}{4}}\right)^{2x}$$

$$= \sum_{k=0}^{\infty} \frac{1}{k!}(e^{\frac{t}{2}}-e^{-\frac{t}{2}})^k x^{[k]}.$$ (6)

Now, it can be easily seen that Equation (4) follows from (5) and (6).

Kim-Kim in [11] introduced the central Bell polynomials by means of generating function as

$$e^{x\left(e^{\frac{t}{2}}-e^{-\frac{t}{2}}\right)} = \sum_{n=0}^{\infty} B_n^{(c)}(x)\frac{t^n}{n!}.$$ (7)

We note by making use of (4) that identity (7) implies (see [1,11])

$$B_n^{(c)}(x) = \sum_{k=0}^{n} T(n,k)x^k, \quad (n \geq 0).$$

For a nonnegative integer r, Kim-Dolgy-Kim-Kim in a recent work [1] introduced the extended r-central factorial numbers of the second kind given by the generating function:

$$\frac{1}{k!}\left(e^{\frac{t}{2}}-e^{-\frac{t}{2}}\right)^k e^{rt} = \sum_{n=k}^{\infty} T^{(r)}(n+r,k+r)\frac{t^n}{n!}.$$ (8)

From (8), it is noted that (see [1])

$$(x+r)^n = \sum_{k=0}^{n} T^{(r)}(n+r,k+r)x^{[k]}.$$ (9)

The extended r-central Bell polynomials [1] are defined by

$$e^{x\left(e^{\frac{t}{2}}-e^{-\frac{t}{2}}\right)} e^{rt} = \sum_{n=0}^{\infty} B_n^{(c,r)}(x)\frac{t^n}{n!}, \quad .$$ (10)

By definition (10), it is also known that (see [1])

$$B_n^{(c,r)}(x) = \sum_{k=0}^{n} x^k T^{(r)}(n+r,k+r), \quad (n \geq 0).$$ (11)

The purpose of this paper is to introduce and study the extended r-central incomplete and complete Bell polynomials, as multivariate versions of the recently studied the extended r-central factorial numbers of the second and the extended r-central Bell polynomials (see [1]), and also as multivariate versions of the r- Stirling numbers of the second kind and the extended r-Bell polynomials (see Section 2). Then we investigate their properties, some identities and various explicit formulas related to these polynomials and also their connections.

This paper is organized as follows. In Section 2, we introduce the incomplete and complete r-Bell polynomials and give some of their simple properties. We observe that these polynomials are multivariate versions of the r- Stirling numbers of the second kind and the extended r-Bell polynomials . In Section 3, we introduce our object of study, namely the extended r-central incomplete and complete Bell polynomials, and provide several properties, some identities and various explicit formulas for them. Finally, in Section 4, brief summaries for the obtained results about newly defined polynomials are provided.

2. Preliminaries

The r-Stirling numbers $S_2^{(r)}(n,k)$ of the second kind are defined by the generating function (see [14–19])

$$\frac{1}{k!}\left(e^t - 1\right)^k e^{rt} = \sum_{n=k}^{\infty} S_2^{(r)}(n+r, k+r)\frac{t^n}{n!} \tag{12}$$

and they enumerate the number of partitions of the set $\{1, 2, \cdots, n\}$ into k nonempty disjoint subsets in such a way that $1, 2, \cdots, r$ are in distinct subsets.

The extended r-Bell polynomials are given by (see [15])

$$e^{rt}e^{x(e^t-1)} = \sum_{n=0}^{\infty} B_n^{(r)}(x)\frac{t^n}{n!}. \tag{13}$$

One can show that Equations (12) and (13) imply

$$\begin{aligned} B_n^{(r)}(x) &= e^{-x}\sum_{k=0}^{\infty}\frac{(k+r)^n}{k!}x^k \\ &= \sum_{k=0}^{n} x^k S_2^{(r)}(n+r, k+r), \quad (n \geq 0). \end{aligned} \tag{14}$$

In particular $x = 1$, $B_n^{(c,r)} = B_n^{(c,r)}(1)$ are called the extended r-Bell numbers.

The incomplete r-Bell polynomials are given by the generating function

$$\frac{1}{k!}\left(\sum_{j=1}^{\infty} x_j \frac{t^j}{j!}\right)^k \left(\sum_{j=0}^{\infty} y_{j+1}\frac{t^j}{j!}\right)^r = \sum_{n \geq k} B_{n+r,k+r}^{(r)}(x_1, x_2, \cdots ; y_1, y_2 \cdots)\frac{t^n}{n!}. \tag{15}$$

Thus, we have

$$\begin{aligned} &B_{n+r,k+r}^{(r)}(x_1, x_2, \cdots ; y_1, y_2 \cdots) \\ &= \sum \left(\frac{n!}{k_1! k_2! \cdots}\left(\frac{x_1}{1!}\right)^{k_1}\left(\frac{x_2}{2!}\right)^{k_2}\cdots\right)\left(\frac{r!}{r_0! r_1! r_2! \cdots}\left(\frac{y_1}{0!}\right)^{r_0}\left(\frac{y_2}{1!}\right)^{r_1}\left(\frac{y_3}{2!}\right)^{r_2}\cdots\right), \end{aligned} \tag{16}$$

where the summation is over all integers $k_1, k_2, \cdots \geq 0$ and $r_0, r_1, r_2 \cdots \geq 0$, such that

$$\sum_{i\geq 1} k_i = k, \quad \sum_{j\geq 0} r_j = r, \quad \text{and} \quad (k_1 + r_1) + 2(k_2 + r_2) + 3(k_3 + r_3) + \cdots = n.$$

Let $a_1, a_2, \cdots$, and $b_1, b_2, \cdots$ be any sequences of nonnegative integers. Then, as was noted in [20], $B_{n+r,k+r}^{(r)}(a_1, a_2, \cdots ; b_1, b_2 \cdots)$ enumerates the number of partitions of a set with $(n + r)$ elements into $(k + r)$ blocks satisfying:

- The first r elements are in different blocks,
- Any block of size i with no elements of the first r elements, can be colored with a_i colors,
- Any block of size i with one element of the first r elements, can be colored with b_i colors.

From (12) and (16), we note that

$$B_{n+r,k+r}^{(r)}(1, 1, \cdots ; 1, 1, \cdots) = S_2^{(r)}(n + k, k + r), \tag{17}$$

$$B_{n+r,k+r}^{(r)}(\alpha x_1, \alpha x_2, \cdots ; \alpha y_1, \alpha y_2 \cdots) = \alpha^{k+r} B_{n+k,k+r}^{(r)}(x_1, x_2, \cdots ; y_1, y_2 \cdots), \tag{18}$$

and

$$B_{n+r,k+r}^{(r)}(\alpha x_1, \alpha^2 x_2, \cdots ; y_1, \alpha y_2, \alpha^2 y_3, \cdots) = \alpha^n B_{n+k,k+r}^{(r)}(x_1, x_2, \cdots ; y_1, y_2 \cdots), \tag{19}$$

where α is a real number.

By using (15), we get

$$
\begin{aligned}
\sum_{n=k}^{\infty} B_{n+r,k+r}^{(r)}(x, 1, 0, 0, \cdots ; 1, 0, 0, \cdots) \frac{t^n}{n!} &= \frac{1}{k!} \left(xt + \frac{t^2}{2} \right)^k \\
&= t^k \frac{1}{k!} \sum_{n=0}^{k} \binom{k}{n} \left(\frac{t}{2} \right)^n x^{k-n} \\
&= \sum_{n=0}^{k} \frac{(n+k)!}{k!} \binom{k}{n} \left(\frac{1}{2} \right)^n x^{k-n} \frac{t^{n+k}}{(n+k)!}.
\end{aligned}
\tag{20}
$$

Also, it can be seen that

$$\sum_{n=k}^{\infty} B_{n+r,k+r}^{(r)}(x, 1, 0, 0, \cdots ; 1, 0, 0, \cdots) \frac{t^n}{n!} = \sum_{n=0}^{\infty} B_{n+k+r,k+r}^{(r)}(x, 1, 0, 0, \cdots ; 1, 0, 0, \cdots) \frac{t^{n+k}}{(n+k)!}. \tag{21}$$

Thus, by (20) and (21), we have the following equation given by

$$B_{n+k+r,k+r}^{(r)}(x, 1, 0, 0, \cdots ; 1, 0, 0, \cdots) = \begin{cases} \frac{(n+k)!}{k!} \binom{k}{n} \left(\frac{1}{2} \right)^n x^{k-n}, & \text{if } 0 \leq n \leq k, \\ 0, & \text{if } n > k. \end{cases} \tag{22}$$

By replacing n by $n - k$ in (22), we get

$$B_{n+r,k+r}^{(r)}(x, 1, 0, 0, \cdots ; 1, 0, 0, \cdots) = \frac{n!}{k!} \binom{k}{n-k} x^{2k-n} \left(\frac{1}{2} \right)^{n-k}, \quad (k \leq n \leq 2k). \tag{23}$$

Now, we define the complete r-Bell polynomials by virtue of generating function as

$$\exp\left(\sum_{i=1}^{\infty} x_i \frac{t^i}{i!}\right)\left(\sum_{j=0}^{\infty} y_{j+1}\frac{t^j}{j!}\right)^r = \sum_{n=0}^{\infty} B_n^{(r)}(x_1, x_2, \cdots ; y_1, y_2, \cdots)\frac{t^n}{n!}. \tag{24}$$

From (15) and (24), we have

$$\begin{aligned}
\sum_{n=0}^{\infty} B_n^{(r)}(x_1, x_2, \cdots ; y_1, y_2, \cdots)\frac{t^n}{n!} &= \sum_{k=0}^{\infty}\frac{1}{k!}\left(\sum_{i=1}^{\infty} x_i\frac{t^i}{i!}\right)^k \left(\sum_{j=0}^{\infty} y_{j+1}\frac{t^j}{j!}\right)^r \\
&= \sum_{k=0}^{\infty}\sum_{n=k}^{\infty} B_{n+r,k+r}^{(r)}(x_1, x_2, \cdots ; y_1, y_2, \cdots)\frac{t^n}{n!} \\
&= \sum_{n=0}^{\infty}\sum_{k=0}^{n} B_{n+r,k+r}^{(r)}(x_1, x_2, \cdots ; y_1, y_2, \cdots)\frac{t^n}{n!}.
\end{aligned} \tag{25}$$

Comparing both sides of (25) gives us the identity

$$B_n^{(r)}(x_1, x_2, \cdots ; y_1, y_2, \cdots) = \sum_{k=0}^{n} B_{n+r,k+r}^{(r)}(x_1, x_2, \cdots ; y_1, y_2, \cdots). \tag{26}$$

Now, we observe that

$$\begin{aligned}
B_n^{(r)}(x, x, \cdots ; 1, 1, \cdots) &= \sum_{k=0}^{n} B_{n+r,k+r}^{(r)}(x, x, \cdots ; 1, 1, \cdots) \\
&= \sum_{k=0}^{n} x^k B_{n+r,k+r}^{(r)}(1, 1, \cdots ; 1, 1, \cdots) \\
&= \sum_{k=0}^{n} x^k S_2^{(r)}(n+r, k+r) \\
&= B_n^{(r)}(x), \quad (n \geq 0).
\end{aligned} \tag{27}$$

3. An Extended r-Central Complete and Incomplete Bell Polynomials

Recently, in [21], we initiated the study of central incomplete Bell polynomials $T_{n,k}(x_1, x_2, \cdots, x_{n-k+1})$ and the central complete Bell polynomials $B_n^{(c)}(x|x_1, x_2, \cdots, x_n)$, respectively given by

$$\frac{1}{k!}\left(\sum_{m=1}^{\infty}\frac{1}{2^m}(x_m - (-1)^m x_m)\frac{t^m}{m!}\right)^k = \sum_{n=k}^{\infty} T_{n,k}(x_1, x_2, \cdots, x_{n-k+1})\frac{t^n}{n!},$$

and

$$\exp\left(x\sum_{i=1}^{\infty}\frac{1}{2^i}(x_i - (-1)^i x_i)\frac{t^i}{i!}\right) = \sum_{n=0}^{\infty} B_n^{(c)}(x|x_1, x_2, \cdots, x_n)\frac{t^n}{n!},$$

and studied some properties and identities concerning these polynomials. It was observed, in particular, that the number of partitioning a set with n elements into k blocks with odd sizes is given by the number of monomials appearing in $T_{n,k}(x_1, 2x_2, \cdots, 2^{n-k}x_{n-k+1})$, and that the number of partitioning a set with n elements into a certain k blocks with odd sizes is the coefficient of the corresponding monomial appearing in $T_{n,k}(x_1, 2x_2, \cdots, 2^{n-k}x_{n-k+1})$.

Here we will consider 'r-extensions' of the central incomplete and complete Bell polynomials. In light of (15), we may define the extended r-central incomplete Bell polynomials by

$$\frac{1}{k!}\left(\sum_{m=1}^{\infty}\left(\frac{1}{2}\right)^m (x_m-(-1)^m x_m)\frac{t^m}{m!}\right)^k \left(\sum_{j=0}^{\infty}y_{j+1}\frac{t^j}{j!}\right)^r = \sum_{n=k}^{\infty}T^{(r)}_{n+r,k+r}(x_1,x_2,\cdots;y_1,y_2,\cdots)\frac{t^n}{n!} \qquad (28)$$

for any $k \in \mathbb{N}\cup\{0\}$. Then, for $n,k \geq 0$ with $n \geq k$, by (28), one can check that

$$\begin{aligned}
T^{(r)}_{n+r,k+r}(x_1,x_2,\cdots;y_1,y_2,\cdots) &= \sum\left(\frac{n!}{k_1!k_3!k_5!\cdots}\left(\frac{x_1}{1!}\right)^{k_1}\left(\frac{x_3}{2^2 3!}\right)^{k_3}\left(\frac{x_5}{2^4 5!}\right)^{k_5}\cdots\right)\\
&\quad \times\left(\frac{r!}{r_0!r_1!r_2!\cdots}\left(\frac{y_1}{0!}\right)^{r_0}\left(\frac{y_2}{1!}\right)^{r_1}\left(\frac{y_3}{2!}\right)^{r_2}\cdots\right),
\end{aligned} \qquad (29)$$

where the summation is over all integers $k_1,k_3,k_5\cdots \geq 0$ and $r_0,r_1,r_2\cdots \geq 0$, such that

$$\sum_{i\geq 1}k_{2i-1}=k, \quad \sum_{i\geq 0}r_i=r, \quad \text{and}\quad \sum_{i\geq 1}(2i-1)k_{2i-1}+\sum_{i\geq 1}ir_i=n. \qquad (30)$$

The extended r-central incomplete Bell polynomials have the following combinatorial interpretation. This can be seen from (29). Let $a_1,a_2,\cdots$, and $b_1,b_2,\cdots$ be any sequences of nonnegative integers. Then $T^{(r)}_{n+r,k+r}(a_1,2a_2,2^2a_3,\cdots;b_1,b_2,b_3,\cdots)$ enumerates the number of partitions of a set with $(n+r)$ elements into k blocks of odd sizes and r blocks of any sizes satisfying:

- The first r elements are in different blocks,
- Any block of (odd) size i with no elements of the first r elements, can be colored with a_i colors,
- Any block of size i with one element of the first r elements, can be colored with b_i colors.

From (15), (16) and (29), we note that

$$T^{(r)}_{n+r,k+r}(x_1,x_2,\cdots;y_1,y_2,\cdots) = B^{(r)}_{n+r,k+r}\left(x_1,0,\frac{x_3}{2^2},0,\cdots;y_1,y_2,y_3\cdots\right). \qquad (31)$$

Therefore, we obtain the following theorem.

Theorem 1. *For $n,k \geq 0$, with $n \geq k$, we have*

$$T^{(r)}_{n+r,k+r}(x_1,x_2,\cdots;y_1,y_2,\cdots) = B^{(r)}_{n+r,k+r}\left(x_1,0,\frac{x_3}{2^2},0,\cdots;y_1,y_2,y_3\cdots\right).$$

From (28), we have

$$\begin{aligned}
\sum_{n=k}^{\infty}T^{(r)}_{n+r,k+r}(1,1,\cdots;1,1,\cdots)\frac{t^n}{n!} &= \frac{1}{k!}\left(\sum_{m=1}^{\infty}\left(\frac{1}{2}\right)^m(1-(-1)^m)\frac{t^m}{m!}\right)^k\left(\sum_{j=0}^{\infty}\frac{t^j}{j!}\right)^r\\
&= \frac{1}{k!}\left(e^{\frac{t}{2}}-e^{-\frac{t}{2}}\right)^k e^{rt}\\
&= \sum_{n=k}^{\infty}T^{(r)}(n+r,k+r)\frac{t^n}{n!}.
\end{aligned} \qquad (32)$$

Therefore, by (32), we obtain the following corollary.

Corollary 1. *For $n, k \geq 0$, with $n \geq k$, we have*

$$T^{(r)}_{n+r,k+r}(1,1,\cdots;1,1,\cdots) = T^{(r)}(n+r,k+r), \quad (r \in \mathbb{N} \cup \{0\}).$$

Let n, k be nonnegative integers. Then, from (28), we get

$$
\begin{aligned}
\sum_{n=k}^{\infty} T^{(r)}_{n+r,k+r}(x, x^2, x^3, \cdots; 1, x, x^2, \cdots)\frac{t^n}{n!} &= \frac{1}{k!}\left(xt + \frac{x^3}{2^2}\frac{t^3}{3!} + \frac{x^5}{2^4}\frac{t^5}{5!} + \cdots\right)^k \left(1 + xt + \frac{x^2}{2}t^2 + \cdots\right)^r \\
&= \frac{1}{k!}\left(e^{\frac{xt}{2}} - e^{-\frac{xt}{2}}\right)^k e^{rxt} \\
&= \frac{1}{k!}\sum_{l=0}^{k}\binom{k}{l}(-1)^{k-l}e^{(l+r-\frac{k}{2})xt} \\
&= \sum_{n=0}^{\infty}\frac{x^n}{k!}\sum_{l=0}^{k}\binom{k}{l}(-1)^{k-l}\left(l+r-\frac{k}{2}\right)^n \frac{t^n}{n!}.
\end{aligned}
$$

$$(33)$$

Therefore, comparing both sides of (33) yields the following theorem.

Theorem 2. *For $n, k \geq 0$, we have*

$$
\frac{x^n}{k!}\sum_{l=0}^{k}\binom{k}{l}(-1)^{k-l}\left(l+r-\frac{k}{2}\right)^n = \begin{cases} T^{(r)}_{n+r,k+r}(x, x^2, x^3, \cdots; 1, x, x^2, \cdots), & \text{if } n \geq k, \\ 0, & \text{otherwise.} \end{cases}
$$

In [10], Kim-Dolgy-Kim-Kim proved the following equation (34) given by

$$
\frac{1}{k!}\sum_{l=0}^{k}\binom{k}{l}(-1)^{k-l}\left(l+r-\frac{k}{2}\right)^n = \begin{cases} T^{(r)}(n+r,k+r), & \text{if } n \geq k, \\ 0, & \text{otherwise,} \end{cases}
$$

$$(34)$$

where $n, k \in \mathbb{Z}$ with $n, k \geq 0$. Therefore, by (34), the following corollary is established.

Corollary 2. *For $n, k \in \mathbb{N} \cup \{0\}$, with $n \geq k$, we have*

$$T^{(r)}_{n+r,k+r}(x, x^2, x^3, \cdots; 1, x, x^2, \cdots) = x^n T^{(r)}(n+r,k+r).$$

From (29) and Corollary 2 , one can also have the following identity.

Corollary 3. *For $n, k \geq 0$, with $n \geq k$, we have*

$$T^{(r)}_{n+r,k+r}(x, x^2, x^3, \cdots; 1, x, x^2, \cdots) = x^n T^{(r)}_{n+r,k+r}(1,1,\cdots;1,1,\cdots)$$

and

$$T^{(r)}_{n+r,k+r}(1,1,\cdots;1,1,\cdots) = T^{(r)}(n+r,k+r)$$

$$= B^{(r)}_{n+r,r}\left(1,0,\frac{1}{2^2},0,\frac{1}{2^4},0,\cdots;1,1,1,\cdots\right)$$

$$= \sum\left(\frac{n!}{k_1!k_3!k_5!\cdots}\left(\frac{1}{1!}\right)^{k_1}\left(\frac{1}{2^23!}\right)^{k_3}\left(\frac{1}{2^45!}\right)^{k_5}\cdots\right)$$

$$\times\left(\frac{r!}{r_0!r_1!r_2!\cdots}\left(\frac{1}{0!}\right)^{r_0}\left(\frac{1}{1!}\right)^{r_1}\left(\frac{1}{2!}\right)^{r_2}\cdots\right),$$

where the summation is over all integers $k_1,k_3,k_5\cdots\geq 0$ *and* $r_0,r_1,r_2\cdots\geq 0$, *satisfying the conditions in* (30).

For $n,\ k\geq 0$, we have

$$\sum_{n=k}^{\infty}T^{(r)}_{n+r,k+r}(x,1,0,0,\cdots;1,0,0,\cdots)\frac{t^n}{n!} = \frac{1}{k!}(xt)^k. \tag{35}$$

By comparing the coefficients on both sides of (35), we have

$$T^{(r)}_{n+r,k+r}(x,1,0,0,\cdots;1,0,0,\cdots) = x^k\binom{0}{n-k}. \tag{36}$$

Also, by (29), one can obtain that

$$T^{(r)}_{n+r,k+r}(x,x,\cdots;y,y,\cdots) = x^k y^r T^{(r)}_{n+r,k+r}(1,1,\cdots;1,1,\cdots)$$
$$= x^k y^r T^{(r)}(n+r,k+r), \tag{37}$$

and

$$T^{(r)}_{n+r,k+r}(\alpha x_1,\alpha x_2,\cdots;\alpha y_1,\alpha y_2,\cdots) = \alpha^{k+r}T^{(r)}_{n+r,k+r}(x_1,x_2,\cdots;y_1,y_2,\cdots), \tag{38}$$

where n,k are nonnegative integers with $n\geq k$.

Now, we observe that

$$\exp\left(x\sum_{i=1}^{\infty}\left(\frac{1}{2}\right)^i\left(x_i-(-1)^ix_i\right)\frac{t^i}{i!}\right)\left(\sum_{j=0}^{\infty}y_{j+1}\frac{t^j}{j!}\right)^r$$

$$= \sum_{k=0}^{\infty}x^k\frac{1}{k!}\left(\sum_{i=1}^{\infty}\left(\frac{1}{2}\right)^i\left(x_i-(-1)^ix_i\right)\frac{t^i}{i!}\right)^k\left(\sum_{j=0}^{\infty}y_{j+1}\frac{t^j}{j!}\right)^r$$

$$= \sum_{k=0}^{\infty}x^k\sum_{n=k}^{\infty}T^{(r)}_{n+r,k+r}(x_1,x_2,\cdots;y_1,y_2,\cdots)\frac{t^n}{n!} \tag{39}$$

$$= \sum_{k=0}^{\infty}\sum_{k=0}^{n}x^k T^{(r)}_{n+r,k+r}(x_1,x_2,\cdots;y_1,y_2,\cdots)\frac{t^n}{n!}.$$

Taking (24) into account, we may define the extended r-central complete Bell polynomials by

$$\exp\left(x\sum_{i=1}^{\infty}\left(\frac{1}{2}\right)^i\left(x_i-(-1)^ix_i\right)\frac{t^i}{i!}\right)\left(\sum_{j=0}^{\infty}y_{j+1}\frac{t^j}{j!}\right)^r = \sum_{k=0}^{\infty}B^{(c,r)}_n(x|x_1,x_2,\cdots;y_1,y_2,\cdots)\frac{t^n}{n!}. \tag{40}$$

In particular, when $x = 1$, $B_n^{(c,r)}(1|x_1, x_2, \cdots ; y_1, y_2, \cdots) = B_n^{(c,r)}(x_1, x_2, \cdots ; y_1, y_2, \cdots)$ are called the extended r-central complete Bell numbers.

For $n \geq 0$, by (39) and (40), we get

$$B_n^{(c,r)}(x_1, x_2, \cdots ; y_1, y_2, \cdots) = \sum_{k=0}^{n} T_{n+r,k+r}^{(r)}(x_1, x_2, \cdots ; y_1, y_2, \cdots) \tag{41}$$

and

$$B_n^{(c,r)}(x|x_1, x_2, \cdots ; y_1, y_2, \cdots) = \sum_{k=0}^{n} x^k T_{n+r,k+r}^{(r)}(x_1, x_2, \cdots ; y_1, y_2, \cdots).$$

It is easily noted that $B_0^{(c,r)}(x_1, x_2, \cdots ; y_1, y_2, \cdots) = y_1^r$.

Hence, one can have the following theorem.

Theorem 3. *For $n \geq 0$, we have*

$$B_n^{(c,r)}(x|x_1, x_2, \cdots ; y_1, y_2, \cdots) = \sum_{k=0}^{n} x^k T_{n+r,k+r}^{(r)}(x_1, x_2, \cdots ; y_1, y_2, \cdots)$$

and

$$B_n^{(c,r)}(x_1, x_2, \cdots ; y_1, y_2, \cdots) = \sum_{k=0}^{n} T_{n+r,k+r}^{(r)}(x_1, x_2, \cdots ; y_1, y_2, \cdots).$$

Please note that

$$B_n^{(c,r)}(1, 1, , \cdots ; 1, 1, \cdots) = \sum_{k=0}^{n} T_{n+r,k+r}^{(r)}(1, 1, \cdots ; 1, 1, \cdots)$$

$$= \sum_{k=0}^{n} T^{(r)}(n+r, k+r)$$

$$= B_n^{(c,r)},$$

and

$$B_n^{(c,r)}(x|1, 1, , \cdots ; 1, 1, \cdots) = \sum_{k=0}^{n} x^k T^{(r)}(n+r, k+r) = B_n^{(c,r)}(x), \quad (n \geq 0).$$

By (39), we get

$$\exp\left(\sum_{i=1}^{\infty} \left(\frac{1}{2}\right)^i \left(x_i - (-1)^i x_i\right) \frac{t^i}{i!} \right) \left(\sum_{j=0}^{\infty} y_{j+1} \frac{t^j}{j!} \right)^r$$

$$= \sum_{n=0}^{\infty} \left\{ \sum \left(\frac{n!}{k_1! k_3! k_5! \cdots} \left(\frac{x_1}{1!}\right)^{k_1} \left(\frac{x_3}{2^2 3!}\right)^{k_3} \left(\frac{x_5}{2^4 5!}\right)^{k_5} \cdots \right) \right. \tag{42}$$

$$\left. \times \left(\frac{r!}{r_0! r_1! r_2! \cdots} \left(\frac{y_1}{0!}\right)^{r_0} \left(\frac{y_2}{1!}\right)^{r_1} \left(\frac{y_3}{2!}\right)^{r_2} \cdots \right) \right\} \frac{t^n}{n!},$$

where the inner sum runs over all integers $k_1, k_3, k_5 \cdots \geq 0$ and $r_0, r_1, r_2 \cdots \geq 0$, such that

$$\sum_{i \geq 0} r_i = r, \text{ and } \sum_{i \geq 1}(2i - 1)k_{2i-1} + \sum_{i \geq 1} i r_i = n. \tag{43}$$

For $n \geq 0$, we have

$$
\begin{aligned}
B_n^{(c,r)}(x_1, x_2, \cdots ; y_1, y_2, \cdots) &= \sum_{k=0}^{n} T_{n+r,k+r}^{(r)}(x_1, x_2, \cdots ; y_1, y_2, \cdots) \\
&= \sum \left(\frac{n!}{k_1! k_3! k_5! \cdots} \left(\frac{x_1}{1!}\right)^{k_1} \left(\frac{x_3}{2^2 3!}\right)^{k_3} \left(\frac{x_5}{2^4 5!}\right)^{k_5} \cdots \right) \\
&\quad \times \left(\frac{r!}{r_0! r_1! r_2! \cdots} \left(\frac{y_1}{0!}\right)^{r_0} \left(\frac{y_2}{1!}\right)^{r_1} \left(\frac{y_3}{2!}\right)^{r_2} \cdots \right),
\end{aligned}
\tag{44}
$$

where the sum is over all integers $k_1, k_3, k_5 \cdots \geq 0$ and $r_0, r_1, r_2 \cdots \geq 0$, satisfying the conditions in (43). Thus, the following theorem is established.

Theorem 4. *For $n \geq 0$, we have*

$$
\begin{aligned}
B_n^{(c,r)}(x_1, x_2, \cdots ; y_1, y_2, \cdots) &= \sum \left(\frac{n!}{k_1! k_3! k_5! \cdots} \left(\frac{x_1}{1!}\right)^{k_1} \left(\frac{x_3}{2^2 3!}\right)^{k_3} \left(\frac{x_5}{2^4 5!}\right)^{k_5} \cdots \right) \\
&\quad \times \left(\frac{r!}{r_0! r_1! r_2! \cdots} \left(\frac{y_1}{0!}\right)^{r_0} \left(\frac{y_2}{1!}\right)^{r_1} \left(\frac{y_3}{2!}\right)^{r_2} \cdots \right),
\end{aligned}
$$

where the sum is over all integers $k_1, k_3, k_5 \cdots \geq 0$ and $r_0, r_1, r_2 \cdots \geq 0$, satisfying the conditions in (43).

Now, we observe that

$$
\begin{aligned}
\exp\left(x \sum_{i=1}^{\infty} \left(\frac{1}{2}\right)^i \left(1 - (-1)^i\right) \frac{t^i}{i!} \right) \left(\sum_{j=0}^{\infty} \frac{t^j}{j!} \right)^r &= \sum_{k=0}^{\infty} x^k \frac{1}{k!} \left(\sum_{i=1}^{\infty} \left(\frac{1}{2}\right)^i \left(1 - (-1)^i\right) \frac{t^i}{i!} \right)^k \left(\sum_{j=0}^{\infty} \frac{t^j}{j!} \right)^r \\
&= \sum_{k=0}^{\infty} x^k \sum_{n=k}^{\infty} T_{n+r,k+r}^{(r)}(1,1,\cdots ; 1,1,\cdots) \frac{t^n}{n!} \\
&= \sum_{n=0}^{\infty} \sum_{k=0}^{n} x^k T_{n+r,k+r}^{(r)}(1,1,\cdots ; 1,1,\cdots) \frac{t^n}{n!}.
\end{aligned}
\tag{45}
$$

Alternatively, the left hand side of (45) can be simplified in the following way:

$$
\exp\left(x \sum_{i=1}^{\infty} \left(\frac{1}{2}\right)^i \left(1 - (-1)^i\right) \frac{t^i}{i!} \right) \left(\sum_{j=0}^{\infty} \frac{t^j}{j!} \right)^r = e^{x\left(e^{\frac{t}{2}} - e^{-\frac{t}{2}}\right)} e^{rt} = \sum_{n=0}^{\infty} B_n^{(c,r)}(x) \frac{t^n}{n!}.
\tag{46}
$$

Comparing the coefficients in (45) and (46) gives the following identity.

Theorem 5. *For $n \geq 0$, we have*

$$
\sum_{k=0}^{n} x^k T_{n+r,k+r}^{(r)}(1,1,\cdots ; 1,1,\cdots) = B_n^{(c,r)}(x).
$$

From (29), it is noted that

$$
\begin{aligned}
\sum_{k=0}^{n} x^{k+r} T_{n+r,k+r}^{(r)}(1,1,\cdots ; 1,1,\cdots) &= \sum_{k=0}^{n} T_{n+r,k+r}^{(r)}(x, x, \cdots ; x, x, \cdots) \\
&= B_n^{(c,r)}(x, x, \cdots ; x, x, \cdots),
\end{aligned}
$$

which yields the next corollary.

Corollary 4. *For $n \geq 0$, we have*

$$B_n^{(c,r)}(x, x, \cdots ; x, x, \cdots) = x^r B_n^{(c,r)}(x).$$

4. Conclusions

In recent years, studies on various old and new special numbers and polynomials have received attention from many mathematicians. They have been carried out by several means, including generating functions, combinatorial methods, umbral calculus, p-adic analysis, differential equations, probability and so on.

In this paper, by making use of generating functions we introduced and studied the extended r-central incomplete and complete Bell polynomials, as multivariate versions of the recently studied the extended r-central factorial numbers of the second and the extended r-central Bell polynomials (see [1]), and also as multivariate versions of the r- Stirling numbers of the second kind and the extended r-Bell polynomials (see Section 2). Then we studied several properties, some identities and various explicit formulas related to these polynomials and also their connections.

In Section 1 we briefly recalled, in more detail, definitions and basic properties about the central factorial numbers of the second kind, the central Bell polynomials, the extended r-central factorial numbers of the second kind and the extended r-central Bell polynomials. In Section 2 we introduced the incomplete and complete r-Bell polynomials as multivariate versions of the r- Stirling numbers of the second kind and the extended r-Bell polynomials and give some of their simple properties. In Section 3, we introduced the extended r-central incomplete and complete Bell polynomials, and provided several properties, some identities and various explicit formulas for them.

As our immediate next project, we would like to find some further results about the extended r-central incomplete and complete Bell polynomials by virtue of umbral calculus and also some of their applications associated with partition polynomials.

Author Contributions: Conceptualization, D.S.K. and T.K.; Formal analysis, D.S.K., D.K. and T.K.; Funding acquisition, D.K.; Investigation, D.S.K., H.Y.K. and T.K.; Methodology, D.S.K. and T.K.; Project administration, T.K.; Software, D.K.; Supervision, D.S.K. and T.K.; Validation, D.S.K., H.Y.K., D.K. and T.K.; Visualization, H.Y.K. and D.K.; Writing—original draft, T.K.; Writing— review & editing, D.S.K., H.Y.K. and D.K.

Funding: This work was supported by the National Research Foundation of Korea (NRF) grant funded by the Korea government (MSIT) (No. 2019R1C1C1003869).

Conflicts of Interest: The authors declare no conflict of interest.

Acknowledgments: We would like to thank the referees for their comments and suggestions which improved the original manuscript greatly.

References

1. Kim, D.S.; Dolgy, D.V.; Kim, D.; Kim, T. Some identities on r-central factorial numbers and r-central Bell polynomials. *arXiv* **2019**, arXiv:1903.11689v1.
2. Riordan, J. *Combinatorial Identities*; John Wiley & Sons, Inc.: New York, NY, USA, 1968.
3. Kim, T.; Kim, D.S.; Jang, G.-W.; Kwon, J. Extended central factorial polynomials of the second kind. *Adv. Differ. Equ.* **2019**, 24. [CrossRef]
4. Kim, T.; Kim, D.S. Degenerate central Bell numbers and polynomials. *Rev. Real Acad. Clenc. Exactas Fis. Nat. Ser. A Mat.* **2019**, 1–7. [CrossRef]
5. Zhang, W. Some identities involving the Euler and the central factorinal numbers. *Fibonacci Quart.* **1998**, *36*, 154–157.
6. Butzer, P.L.; Schmidt, M.; Stark, E.L.; Vogt, L. Central factorial numbers; their main properties and some applications. *Numer. Funct. Anal. Optim.* **1989**, *10*, 419–488. [CrossRef]

7. Carlitz, L.; Riordan, J. The divided central differences of zero. *Canad. J. Math.* **1963**, *15*, 94–100. [CrossRef]
8. Carlitz, L. Some remarks on the Bell numbers. *Fibonacci Quart.* **1980** , *18*, 66–73.
9. Charalambides, C.A. Central factorial numbers and related expansions. *Fibonacci Quart.* **1981**, *19*, 451–456.
10. Kim, T. A note on central factorial numbers. *Proc. Jangjeon Math. Soc.* **2018**, *21*, 575–588.
11. Kim, T.; Kim, D.S. A note on central Bell numbers and polynomials. *Adv. Stud. Contemp. Math. (Kyungshang)* **2019**, *27*, 289–298.
12. Belbachir, H.; Djemmada, Y. On central Fubini-like numbers and polynomials. *arXiv* **2018**, arXiv:1811.06734v1.
13. Krzywonos, N.; Alayont, F. Rook polynomials in higher dimensions. *Stud. Summer Sch.* **2009**, *29*; Available online: https://scholarworks.gvsu.edu/sss/29/ (accessed on 2 March 2019).
14. Duran, U.; Acikgoz, M.; Araci, S. On (q, r, w)-Stirling numbers of the second kind. *J. Inequal. Spec. Funct.* **2018**, *9*, 9–16.
15. Kim, T.; Yao, Y.; Kim, D.S.; Jang, G.-W. Degenerate r-Stirling numbers and r-Bell polynomials. *Russ. J. Math. Phys.* **2018**, *25*, 44–58. [CrossRef]
16. Pyo, S.-S. Degenerate Cauchy numbers and polynomials of the fourth kind. *Adv. Stud. Contemp. Math. (Kyungshang)* **2018**, *28*, 127–138.
17. Roman, S. The umbral calculus. In *Pure and Applied Mathematics*; Harcourt Brace Jovanovich: New York, NY, USA, 1984.
18. Simsek, Y. Identities and relations related to combinatorial numbers and polynomials. *Proc. Jangjeon Math. Soc.* **2017**, *20*, 127–135.
19. Simsek, Y. Identities on the Changhee numbers and Apostol-type Daehee polynomials. *Adv. Stud. Contemp. Math. (Kyungshang)* **2017**, *27*, 199–212.
20. Mihoubi, M.; Rahmani, M. The partial r-Bell polynomials. *Afr. Mat.* **2017**, *28*, 1167–1183. [CrossRef]
21. Kim, T.; Kim, D.S.; Jang, G.-W. On central complete and incomplete Bell polynomials I. *Symmetry* **2019**, *11*, 288. [CrossRef]

 *symmetry*

Article

Some Identities of Fully Degenerate Bernoulli Polynomials Associated with Degenerate Bernstein Polynomials

Jeong Gon Lee [1], Wonjoo Kim [2],* and Lee-Chae Jang [3]

[1]   Division of Applied Mathematics, Nanoscale Science and Technology Institute, Wonkwang University, Iksan 54538, Korea; jukolee@wku.ac.kr
[2]   Department of Applied Mathematics, Kyunghee University, Yongin 17104, Korea
[3]   Graduate School of Education, Konkuk University, Seoul 05029, Korea; lcjang@konkuk.ac.kr
*   Correspondence: wjookim@khu.ac.kr

Received: 26 April 2019; Accepted: 20 May 2019; Published: 24 May 2019

Abstract: In this paper, we investigate some properties and identities for fully degenerate Bernoulli polynomials in connection with degenerate Bernstein polynomials by means of bosonic p-adic integrals on $\mathbb{Z}_p$ and generating functions. Furthermore, we study two variable degenerate Bernstein polynomials and the degenerate Bernstein operators.

Keywords: degenerate Bernoulli polynomials; degenerate Bernstein operators

1. Introduction

Let p be a fixed prime number. Throughout this paper, $\mathbb{Z}$, $\mathbb{Z}_p$, $\mathbb{Q}_p$ and $\mathbb{C}_p$, will denote the ring of rational integers, the ring of p-adic integers, the field of p-adic rational numbers and the completion of algebraic closure of $\mathbb{Q}_p$, respectively. The p-adic norm $|q|_p$ is normalized as $|p|_p = \frac{1}{p}$.

For $\lambda, t \in \mathbb{C}_p$ with $|\lambda t|_p < p^{-\frac{1}{p-1}}$ and $|t|_p < 1$, the degenerate Bernoulli polynomials are defined by the generating function to be

$$\frac{t}{(1+\lambda t)^{\frac{1}{\lambda}} - 1}(1+\lambda t)^{\frac{x}{\lambda}} = \sum_{n=0}^{\infty} \beta_n(x|\lambda)\frac{t^n}{n!}, \tag{1}$$

(See [1–3]). When $x = 0$, $\beta_n(\lambda) = \beta_n(0|\lambda)$ are called the degenerate Bernoulli numbers. Note that $\lim_{\lambda \to 0} \beta_n(x|\lambda) = B_n(x)$, where $B_n(x)$ are the ordinary Bernoulli polynomials defined by

$$\frac{t}{e^t - 1}e^{xt} = \sum_{n=0}^{\infty} B_n(x)\frac{t^n}{n!}, \tag{2}$$

and $B_n = B_n(0)$ are called the Bernoulli numbers. The degenerate exponential function is defined by

$$e_\lambda^x(t) = (1+\lambda t)^{\frac{x}{\lambda}} = \sum_{n=0}^{\infty} (x)_{n,\lambda}\frac{t^n}{n!}, \tag{3}$$

where $(x)_{0,\lambda} = 1$, $(x)_{n,\lambda} = x(x-\lambda)(x-2\lambda)\cdots(x-(n-1)\lambda)$, for $n \geq 1$. From (1), we get

$$\beta_n(x|\lambda) = \sum_{l=0}^{n}\binom{n}{l}\beta_l(\lambda)(x)_{n-l,\lambda}. \tag{4}$$

Recentely, Kim-Kim introduced the degenerate Bernstein polynomials given by

$$\frac{(x)_{k,\lambda}}{k!} t^k (1+\lambda t)^{\frac{1-x}{\lambda}} = \sum_{n=k}^{\infty} B_{k,n}(x|\lambda) \frac{t^n}{n!}, \tag{5}$$

(See [4–6]). Thus, by (5), we note that

$$B_{k,n}(x|\lambda) = \begin{cases} \binom{n}{k}(x)_{k,\lambda}(1-x)_{n-k,\lambda}, & \text{if } n \geq k, \\ 0, & \text{if } n < k. \end{cases} \tag{6}$$

where n, k are non-negative integers. Let $UD(\mathbb{Z}_p)$ be the space of uniformly differentiable functions on $\mathbb{Z}_p$. For $f \in UD(\mathbb{Z}_p)$, the degenerate Bernstein operator of order n is given by

$$\begin{aligned}
\mathbb{B}_{n,\lambda}(f|\lambda) &= \sum_{k=0}^{\infty} f\left(\frac{k}{n}\right) \binom{n}{k}(x)_{k,\lambda}(1-x)_{n-k,\lambda} \\
&= \sum_{k=0}^{\infty} f\left(\frac{k}{n}\right) B_{k,n}(x|\lambda),
\end{aligned} \tag{7}$$

(See [4–6]). The bosonic p-adic integral on $\mathbb{Z}_p$ is defined by Volkenborn as

$$\int_{\mathbb{Z}_p} f(x)d\mu_0(x) = \lim_{N \to \infty} \frac{1}{p^N} \sum_{x=0}^{p^N-1} f(x), \tag{8}$$

(see [7]). By (8), we get

$$\int_{\mathbb{Z}_p} f(x+1)d\mu_0(x) - \int_{\mathbb{Z}_p} f(x)d\mu_0(x) = f'(0), \tag{9}$$

where $\frac{d}{dx}f(x)\big|_{x=0} = f'(0)$.

From (8), Kim-Seo [8] proposed fully degenerate Bernoulli polynomials which are reformulated in terms of bosonic p-adic integral on $\mathbb{Z}_p$ as

$$\int_{\mathbb{Z}_p} (1+\lambda t)^{\frac{x+y}{\lambda}} d\mu_0(y) = \frac{\frac{1}{\lambda}\log(1+\lambda t)}{(1+\lambda t)^{\frac{1}{\lambda}} - 1}(1+\lambda t)^{\frac{x}{\lambda}} = \sum_{n=0}^{\infty} B_n(x|\lambda)\frac{t^n}{n!}, \tag{10}$$

and for $x = 0$, $B_n(\lambda) = B_n(0|\lambda)$ are called fully degenerate Bernoulli numbers.

Note that the fully degenerate Bernoulli polynomial was named Daehee polynomials with α-parameter in [9]. On the other hand,

$$\int_{\mathbb{Z}_p} (1+\lambda t)^{\frac{x+y}{\lambda}} d\mu_0(y) = \sum_{n=0}^{\infty} \int_{\mathbb{Z}_p} (x+y)_{n,\lambda} d\mu_0(y)\frac{t^n}{n!}. \tag{11}$$

By (10) and (11), we get

$$\int_{\mathbb{Z}_p} (x+y)_{n,\lambda} d\mu_0(y) = B_n(x|\lambda), \quad (n \geq 0). \tag{12}$$

Recall that the Daehee polynomials are defined by the generating function to be

$$\frac{\log(1+t)}{t}(1+t)^x = \sum_{n=0}^{\infty} D_n(x)\frac{t^n}{n!}, \tag{13}$$

and for $x = 0$, $D_n = D_n(0)$ are called the Daehee numbers (see [10,11]).

Also, the higher order Daehee polynomials are defined by the generating function to be

$$\left(\frac{\log(1+t)}{t}\right)^k (1+t)^x = \sum_{n=0}^{\infty} D_n^{(k)}(x)\frac{t^n}{n!}, \tag{14}$$

and for $x = 0$, $D_n^{(k)} = D_n^{(k)}(0)$ are called the higher order Daehee numbers. From (10), we observe

$$\begin{aligned}
\frac{\frac{1}{\lambda}\log(1+\lambda t)}{(1+\lambda t)^{\frac{1}{\lambda}} - 1}(1+\lambda t)^{\frac{x}{\lambda}} &= \frac{t}{(1+\lambda t)^{\frac{1}{\lambda}} - 1}(1+\lambda t)^{\frac{x}{\lambda}}\frac{\log(1+\lambda t)}{\lambda t} \\
&= \left(\sum_{m=0}^{\infty} \beta_m(x|\lambda)\frac{t^m}{m!}\right)\left(\sum_{l=0}^{\infty} D_l\frac{(\lambda t)^l}{l!}\right) \\
&= \sum_{n=0}^{\infty}\left(\sum_{m=0}^{n}\binom{n}{m}\beta_m(x|\lambda)D_{n-m}\lambda^{n-m}\right)\frac{t^n}{n!}.
\end{aligned} \tag{15}$$

By (10) and (14), we get

$$B_n(x|\lambda) = \sum_{m=0}^{n}\binom{n}{m}\beta_m(x|\lambda)D_{n-m}\lambda^{n-m}, \quad (n \geq 0). \tag{16}$$

From (3) and (10), we observe that

$$\begin{aligned}
\sum_{n=0}^{\infty} B_n(x|\lambda)\frac{t^n}{n!} &= \frac{\frac{1}{\lambda}\log(1+\lambda t)}{(1+\lambda t)^{\frac{1}{\lambda}} - 1}(1+\lambda t)^{\frac{x}{\lambda}} \\
&= \left(\sum_{m=0}^{\infty} B_m(\lambda)\frac{t^m}{m!}\right)\left(\sum_{l=0}^{\infty}(x)_{l,\lambda}\frac{t^l}{l!}\right) \\
&= \sum_{n=0}^{\infty}\left(\sum_{m=0}^{n}\binom{n}{m}B_m(\lambda)(x)_{n-m,\lambda}\right)\frac{t^n}{n!}.
\end{aligned} \tag{17}$$

By (17), we get

$$B_n(x|\lambda) = \sum_{m=0}^{n}\binom{n}{m}B_m(\lambda)(x)_{n-m,\lambda}, \quad (n \geq 0). \tag{18}$$

From (1) and (3), we note that

$$\begin{aligned}
t &= \left((1+\lambda t)^{\frac{1}{\lambda}} - 1\right)\sum_{m=0}^{\infty}\beta_m(\lambda)\frac{t^m}{m!} \\
&= \left(\sum_{l=0}^{\infty}(1)_{l,\lambda}\frac{t^l}{l!}\right)\left(\sum_{m=0}^{\infty}\beta_m(\lambda)\frac{t^m}{m!}\right) - \sum_{m=0}^{\infty}\beta_m(\lambda)\frac{t^m}{m!} \\
&= \sum_{n=0}^{\infty}\left(\sum_{m=0}^{n}\binom{n}{m}(1)_{n-m,\lambda}\beta_m(\lambda)\right)\frac{t^n}{n!} - \sum_{n=0}^{\infty}\beta_n(\lambda)\frac{t^n}{n!} \\
&= \sum_{n=0}^{\infty}\left(\sum_{m=0}^{n}\binom{n}{m}(1)_{n-m,\lambda}\beta_m(\lambda) - \beta_n(\lambda)\right)\frac{t^n}{n!}.
\end{aligned} \tag{19}$$

Comparing the cofficients on both sides of (19), we get

$$\sum_{m=0}^{n}\binom{n}{m}(1)_{n-m,\lambda}\beta_m(\lambda) - \beta_n(\lambda) = \delta_{1,n}, \quad (n \geq 0), \tag{20}$$

where $\delta_{k,n}$ is the Kronecker's symbol.

By (4) and (20), we have

$$\beta_n(1|\lambda) - \beta_n(\lambda) = \delta_{1,n}. \tag{21}$$

The generating function of fully degenerate Bernoulli polynomials introduced in (5) can be expressed as bosonic p-adic integral but the generating function of degenerate Bernoulli polynomials introduced in (1) is not expressed as a bosonic p-adic integral. This is why we considered the fully degenerate Bernoulli polynomials, and the motivation of this paper is to investigate some identities of them associated with degenerate Bernstein polynomials.

In this paper, we consider the fully degenerate Bernoulli polynomials and investigate some properties and identities for these polynomials in connection with degenerate Bernstein polynomials by means of bosonic p-adic integrals on $\mathbb{Z}_p$ and generating functions. Furthermore, we study two variable degenerate Bernstein polynomials and the degenerate Bernstein operators.

2. Fully Degenerate Bernoulli and Bernstein Polynomials

From (10), we observe that

$$
\begin{aligned}
\sum_{n=0}^{\infty} B_n(1-x|\lambda)\frac{t^n}{n!} &= \frac{\frac{1}{\lambda}\log(1+\lambda t)}{(1+\lambda t)^{\frac{1}{\lambda}}-1}(1+\lambda t)^{\frac{1-x}{\lambda}} \\
&= \frac{\left(-\frac{1}{\lambda}\right)(1+(-\lambda)(-t))}{(1+(-\lambda)(-t))^{-\frac{1}{\lambda}}-1}(1+(-\lambda)(-t))^{-\frac{x}{\lambda}} \\
&= \sum_{n=0}^{\infty} B_n(x|-\lambda)(-1)^n\frac{t^n}{n!}.
\end{aligned}
\tag{22}
$$

From (22), we obtain the following Lemma.

Lemma 1. *For $n \in \mathbb{N} \cup \{0\}$, we have*

$$B_n(1-x|\lambda) = (-1)^n B_n(x|-\lambda). \tag{23}$$

From (16) and (21), we get

$$
\begin{aligned}
B_n(1|\lambda) - B_n(\lambda) &= \sum_{m=0}^{n} \binom{n}{m} (\beta_m(1|\lambda) - \beta_m(\lambda)) D_{n-m}\lambda^{n-m} \\
&= \sum_{m=0}^{n} \binom{n}{m} \delta_{1,m} D_{n-m}\lambda^{n-m}, \quad (n \geq 0).
\end{aligned}
\tag{24}
$$

From (1), we observe that

$$
\begin{aligned}
\sum_{n=0}^{\infty} \beta_n(x+1|\lambda)\frac{t^n}{n!} &= (1+\lambda t)^{\frac{1}{\lambda}}\left(\sum_{m=0}^{\infty} \beta_m(x|\lambda)\frac{t^m}{m!}\right) \\
&= \left(\sum_{l=0}^{\infty}(1)_{l,\lambda}\frac{t^l}{l!}\right)\left(\sum_{m=0}^{\infty}\beta_m(x|\lambda)\frac{t^m}{m!}\right) \\
&= \sum_{n=0}^{\infty}\left(\sum_{m=0}^{n}\binom{n}{m}(1)_{n-m,\lambda}\beta_m(x|\lambda)\right)\frac{t^n}{n!}.
\end{aligned}
\tag{25}
$$

By (25), we get

$$\beta_n(x+1|\lambda) = \sum_{m=0}^{n}\binom{n}{m}(1)_{n-m,\lambda}\beta_m(x|\lambda). \tag{26}$$

By (26), with $x = 1$, we have

$$
\begin{aligned}
\beta_n(2|\lambda) &= \sum_{m=0}^{n} \binom{n}{m} \beta_m(1|\lambda)(1)_{n-m,\lambda} \\
&= (1)_{n,\lambda}\beta_0(1|\lambda) + n(1)_{n-1,\lambda}\beta_1(1|\lambda) + \sum_{m=2}^{n} \binom{n}{m} \beta_m(1|\lambda)(1)_{n-m,\lambda} \\
&= (1)_{n,\lambda} + n(1)_{n-1,\lambda}\left(\beta_1(\lambda) - 1\right) + \sum_{m=2}^{n} \binom{n}{m} \beta_m(\lambda)(1)_{n-m,\lambda} \\
&= (1)_{n,\lambda} + n(1)_{n-1,\lambda}\beta_1(\lambda) - n(1)_{n-1,\lambda} + \sum_{m=2}^{n} \binom{n}{m} \beta_m(\lambda)(1)_{n-m,\lambda} \\
&= -n(1)_{n-1,\lambda} + \sum_{m=0}^{n} \binom{n}{m} \beta_m(\lambda)(1)_{n-m,\lambda} \\
&= -n(1)_{n-1,\lambda} + \beta_n(1|\lambda).
\end{aligned}
\tag{27}
$$

Therefore, by (27), we obtain the following theorem.

Theorem 1. *For $n \in \mathbb{N}$, we have*

$$
\beta_n(2|\lambda) = -n(1)_{n-1,\lambda} + \beta_n(1|\lambda).
\tag{28}
$$

Note that

$$
(1-x)_{n,\lambda} = (-1)^n (x-1)_{n,-\lambda}, \quad (n \geq 0).
\tag{29}
$$

Therefore by (12), (23), and (29), we get

$$
\int_{\mathbb{Z}_p} (1-x)_{n,\lambda}\, d\mu_0(x) = (-1)^n \int_{\mathbb{Z}_p} (x-1)_{n,-\lambda}\, d\mu_0(x) = \int_{\mathbb{Z}_p} (x+2)_{n,\lambda}\, d\mu_0(x).
\tag{30}
$$

Therefore, by (30) and Theorem 1, we obtain the following theorem.

Theorem 2. *For $n \in \mathbb{N}$, we have*

$$
\int_{\mathbb{Z}_p} (1-x)_{n,\lambda}\, d\mu_0(x) = \int_{\mathbb{Z}_p} (x+2)_{n,\lambda}\, d\mu_0(x) = n(1)_{n-1,\lambda}(\lambda - 1)B_1(\lambda) + \int_{\mathbb{Z}_p} (x)_{n,\lambda}\, d\mu_0(x).
\tag{31}
$$

Corollary 1. *For $n \in \mathbb{N}$, we have*

$$
(-1)^n B_n(-1|-\lambda) = (1)_{n-1,\lambda}\left(1 - nB_1(\lambda)\right) + B_n(\lambda) = B_n(2|\lambda).
\tag{32}
$$

By (17), we get

$$
\begin{aligned}
B_n(1-x|\lambda) &= \sum_{m=0}^{n} \binom{n}{m} B_m(\lambda)(1-x)_{n-m,\lambda} \\
&= \sum_{m=0}^{n} \binom{n}{m} (x)_{m,\lambda}(1-x)_{n-m,\lambda} \frac{B_m(\lambda)}{(x)_{m,\lambda}} \\
&= \sum_{m=0}^{n} B_{m,n}(x|\lambda)B_m(\lambda)\frac{1}{(x)_{m,\lambda}}.
\end{aligned}
\tag{33}
$$

In [8], we note that

$$\frac{1}{(x)_{m,\lambda}} = \frac{1}{x(x-\lambda)(x-2\lambda)\cdots(x-(m-1)\lambda)}$$
$$= \sum_{k=0}^{m-1} \frac{(-1)^k}{(m-1)!}\binom{m-1}{k}\frac{(-\lambda)^{1-m}}{x-k\lambda}, \quad (m \in \mathbb{N}). \tag{34}$$

By (33) and (34) we get

$$B_n(1-x|\lambda) = \sum_{m=0}^{n} B_{m,n}(x|\lambda)B_m(\lambda)\frac{1}{(x)_{m,\lambda}}$$
$$= (1-x)_{n,\lambda} + \sum_{m=1}^{n} B_{m,n}(x|\lambda)B_m(\lambda)\frac{1}{(x)_{m,\lambda}} \tag{35}$$
$$= (1-x)_{n,\lambda} + \sum_{m=1}^{n} B_{m,n}(x|\lambda)B_m(\lambda)\frac{(-\lambda)^{1-m}}{(m-1)!}\sum_{k=0}^{m-1}(-1)^k\binom{m-1}{k}\frac{1}{x-k\lambda}.$$

Therefore, by (35), we obtain the following theorem.

Theorem 3. *For $n \in \mathbb{N} \cup \{0\}$, we have*

$$B_n(1-x|\lambda) = (1-x)_{n,\lambda} + \sum_{m=1}^{n} B_{m,n}(x|\lambda)B_m(\lambda)\frac{(-\lambda)^{1-m}}{(m-1)!}\sum_{k=0}^{m-1}(-1)^k\binom{m-1}{k}\frac{1}{x-k\lambda}. \tag{36}$$

Corollary 2. *For $n \in \mathbb{N} \cup \{0\}$, we have*

$$B_n(2|\lambda) = (2)_{n,\lambda} + \sum_{m=1}^{n} B_{m,n}(-1|\lambda)B_m(\lambda)\frac{(-\lambda)^{1-m}}{(m-1)!}\sum_{k=0}^{m-1}(-1)^{k+1}\binom{m-1}{k}\frac{1}{1+k\lambda}. \tag{37}$$

For $k \in \mathbb{N}$, the higher-order fully degenerate Bernoulli polynomials are given by the generating function

$$\left(\frac{\frac{1}{\lambda}\log(1+\lambda t)}{(1+\lambda t)^{\frac{1}{\lambda}}-1}\right)^k (1+\lambda t)^{\frac{x}{\lambda}} = \sum_{n=0}^{\infty} B_n^{(k)}(x|\lambda)\frac{t^n}{n!}, \tag{38}$$

(See [8,12,13]). When $x = 0$, $B_n^{(k)}(\lambda) = B_n^{(k)}(x|0)$ are called the higher-order fully degenerate Bernoulli numbers. From (5) and (38), we note that

$$\left(\frac{\log(1+\lambda t)}{\lambda t}\right)^k \sum_{n=k}^{\infty} B_{k,n}(x|\lambda)\frac{t^n}{n!} = (x)_{k,\lambda}t^k(1+\lambda t)^{\frac{1-x}{\lambda}}\left(\frac{\log(1+\lambda t)}{\lambda t}\right)^k\frac{1}{k!}$$
$$= \frac{\left((1+\lambda t)^{\frac{1}{\lambda}}-1\right)^k}{\left((1+\lambda t)^{\frac{1}{\lambda}}-1\right)^k}(x)_{k,\lambda}\left(\frac{1}{\lambda}\log(1+\lambda t)\right)^k(1+\lambda t)^{\frac{1-x}{\lambda}}\frac{1}{k!}$$
$$= (x)_{k,\lambda}\sum_{m=0}^{k}\binom{k}{m}(-1)^{m-k}(1+\lambda t)^{\frac{m}{\lambda}}\left(\frac{\frac{1}{\lambda}\log(1+\lambda t)}{(1+\lambda t)^{\frac{1}{\lambda}}-1}\right)^k(1+\lambda t)^{\frac{1-x}{\lambda}}\frac{1}{k!} \tag{39}$$
$$= (x)_{k,\lambda}\sum_{m=0}^{k}\binom{k}{m}(-1)^{m-k}\left(\frac{\frac{1}{\lambda}\log(1+\lambda t)}{(1+\lambda t)^{\frac{1}{\lambda}}-1}\right)^k(1+\lambda t)^{\frac{1-x+m}{\lambda}}\frac{1}{k!}$$
$$= (x)_{k,\lambda}\sum_{m=0}^{k}\binom{k}{m}(-1)^{m-k}\sum_{n=0}^{\infty}B_n(1-x+m|\lambda)\frac{t^n}{n!}\frac{1}{k!}$$
$$= \sum_{n=0}^{\infty}\left((x)_{k,\lambda}\sum_{m=0}^{k}\binom{k}{m}(-1)^{m-k}B_n(1-x+m|\lambda)\frac{1}{k!}\right)\frac{t^n}{n!},$$

and hence, we get

$$\left(\frac{\log(1+\lambda t)}{\lambda t}\right)^k \sum_{m=k}^{\infty} B_{k,m}(x|\lambda)\frac{t^m}{m!} = \left(\sum_{l=0}^{\infty} D_l^{(k)}\lambda^l\frac{t^l}{l!}\right)\left(\sum_{m=k}^{\infty} B_{k,m}(x|\lambda)\frac{t^m}{m!}\right)$$

$$= \sum_{n=k}^{\infty}\left(\sum_{l=0}^{n} D_l^{(k)}\lambda^l B_{k,n-l}(x|\lambda)\right)\frac{t^n}{n!}. \tag{40}$$

Therefore, by (39) and (40), we obtain the following theorem.

Theorem 4. *For $k, n \in \mathbb{N}$, we have*

$$\frac{1}{k!}(x)_{n,\lambda}\sum_{m=0}^{k}\binom{k}{m}(-1)^{m-k}B_n(1-x+m|\lambda) = \begin{cases}\sum_{l=0}^{n} D_l^{(k)}\lambda^l B_{k,n-l}(x|\lambda), & \text{if } n \geq k, \\ 0, & \text{if } n < k.\end{cases} \tag{41}$$

Let $f \in UD(\mathbb{Z}_p)$. For $x_1, x_2 \in \mathbb{Z}_p$, we consider the degenerate Bernstein operator of order n given by

$$\mathbb{B}_{n,\lambda}(f|x_1,x_2) = \sum_{k=0}^{n} f\left(\frac{k}{n}\right)\binom{n}{k}(x_1)_{k,\lambda}(1-x_2)_{n-k,\lambda} = \sum_{k=0}^{n} f\left(\frac{k}{n}\right) B_{k,n}(x_1,x_2|\lambda), \tag{42}$$

where $B_{n,k}(x_1,x_2|\lambda)$ are called two variable degenerate Bernstein polynomials of degree n as followings (see, [2–6,9,14–27]):

$$B_{k,n}(x_1,x_2|\lambda) = \binom{n}{k}(x_1)_{k,\lambda}(1-x_2)_{n-k,\lambda}, \quad (n \geq 0). \tag{43}$$

The authors [3] obtained the following:

$$\sum_{k=0}^{\infty} B_{k,n}(x_1,x_2|\lambda)\frac{t^n}{n!} = \frac{(x_1)_{k,\lambda}}{k!}t^k e_\lambda^{1-x_2}(t). \tag{44}$$

The authors [8] obtained the following:

$$B_{k,n}(x_1,x_2|\lambda) = \binom{n}{k}(1-(1-x_1))_{n-(n-k),\lambda}(1-x_2)_{n-k,\lambda}$$

$$= B_{n-k,n}(1-x_2,1-x_1|\lambda), \tag{45}$$

and

$$B_{k,n}(x_1,x_2|\lambda) = (1-x_2-(n-k-1)\lambda)B_{k,n-1}(x_1,x_2|\lambda)$$

$$+ (x_1-(k-1)\lambda)B_{k-1,n-1}(x_1,x_2|\lambda). \tag{46}$$

From (42), we note that $x_1, x_2 \in \mathbb{Z}_p$, if $f(x) = 1$, then we have

$$\mathbb{B}_{n,\lambda}(1|x_1,x_2) = \sum_{k=0}^{n} B_{k,n}(x_1,x_2|\lambda)$$

$$= \sum_{k=0}^{n}\binom{n}{k}(x_1)_{k,\lambda}(1-x_2)_{n-k,\lambda} \tag{47}$$

$$= (1+x_1-x_2)_{n,\lambda},$$

and if $f(t) = t$, then we have

$$\mathbb{B}_{n,\lambda}(t|x_1,x_2) = (x_1)_{1,\lambda}(x_1+1-\lambda-x_2)_{n-1,\lambda}, \tag{48}$$

and if $f(t) = t^2$, then we have

$$\mathbb{B}_{n,\lambda}(t^2|x_1,x_2) = \frac{1}{n}(x_1)_{1,\lambda}(x_1+1-\lambda-x_2)_{n-1,\lambda} + \frac{n-1}{n}(x_1)_{2,\lambda}(1+x_2-2\lambda-x_2)_{n-2,\lambda}. \tag{49}$$

The authors [3] obtained the following:

$$(x)_{1,\lambda} = \frac{1}{(x_1+1-\lambda-x_2)_{n-1,\lambda}}\mathbb{B}_n(t|x_1,x_2), \tag{50}$$

and

$$(x)_{2,\lambda} = \frac{1}{(x_1+1-2\lambda-x_2)_{n-2,\lambda}}\sum_{k=2}^{n}\frac{\binom{k}{2}}{\binom{n}{2}}B_{k,n}(x_1,x_2|\lambda), \tag{51}$$

and

$$(x)_{i,\lambda} = \frac{1}{(1+x_1-x_2-i\lambda)_{n-i,\lambda}}\sum_{k=i}^{n}\frac{\binom{k}{i}}{\binom{n}{i}}B_{k,n}(x_1,x_2|\lambda). \tag{52}$$

Taking double bosonic p-adic integral on $\mathbb{Z}_p$, we get the following equation:

$$\int_{\mathbb{Z}_p}\int_{\mathbb{Z}_p}B_{k,n}(x_1,x_2|\lambda)d\mu_0(x_1)d\mu_0(x_2) = \binom{n}{k}\int_{\mathbb{Z}_p}(x_1)_{k,\lambda}d\mu_0(x_1)\int_{\mathbb{Z}_p}(1-x_2)_{n-k,\lambda}d\mu_0(x_2). \tag{53}$$

Therefore, by (53) and Theorem 2, we obtain the following theorem.

Theorem 5. *For $n,k \in \mathbb{N}\cup\{0\}$, we have*

$$\int_{\mathbb{Z}_p}\int_{\mathbb{Z}_p}B_{k,n}(x_1,x_2|\lambda)d\mu_0(x_1)d\mu_0(x_2)$$
$$= \begin{cases} \binom{n}{k}B_n(\lambda)\left((1)_{n-1,\lambda}n(\lambda-1)B_n(\lambda)+B_{n-k}(\lambda)\right), & \text{if } n > k, \\ B_n(\lambda), & \text{if } n = k. \end{cases} \tag{54}$$

We get from the symmetric properties of two variable degenerate Bernstein polynomials that for $n,k \in \mathbb{N}$ with $n > k$,

$$\int_{\mathbb{Z}_p}\int_{\mathbb{Z}_p}B_{k,n}(x_1,x_2|\lambda)d\mu_0(x_1)d\mu_0(x_2)$$
$$= \sum_{m=0}^{k}\binom{n}{k}\binom{k}{m}(-1)^{k+m}(1)_{m,\lambda}$$
$$\qquad \times \int_{\mathbb{Z}_p}\int_{\mathbb{Z}_p}(1-x_1)_{k-m,-\lambda}(1-x_2)_{n-k,\lambda}d\mu_0(x_1)d\mu_0(x_2)$$
$$= \binom{n}{k}\int_{\mathbb{Z}_p}(1-x_2)_{n-k}d\mu_0(x_2)\sum_{m=0}^{k}\binom{n}{k}\binom{k}{m}(-1)^{k+m}(1)_{m,\lambda}d\mu_0(x_2) \tag{55}$$
$$\qquad \times \left\{(1)_{k-m,-\lambda}(k-m)(-\lambda-1)B_1(-\lambda)+\int_{\mathbb{Z}_p}(x_1)_{k-m,-\lambda}d\mu_0(x_1)\right\}$$
$$= \binom{n}{k}B_{n-k,\lambda}(2)\sum_{m=0}^{k}\binom{n}{k}\binom{k}{m}(-1)^{k+m}(1)_{m,\lambda}$$
$$\qquad \times \left\{(1)_{k-m,-\lambda}(k-m)(-\lambda-1)B_1(-\lambda)+B_{k-m,-\lambda}(2)\right\}$$

Therefore, by Theorem 5, we obtain the following theorem.

Theorem 6. *For* $n, k \in \mathbb{N} \cup \{0\}$, *we have the following identities:*

1. *If* $n > k$, *then we have*

$$B_n \left((1)_{n-1,\lambda} n(\lambda - 1) B_1(\lambda) + B_{n-k}(\lambda) \right)$$

$$= B_{n-k,\lambda}(2) \sum_{m=0}^{k} \binom{n}{k} \binom{k}{m} (-1)^{k+m} (1)_{m,\lambda} \tag{56}$$

$$\times \left((1)_{k-m,-\lambda}(k-m)(-\lambda - 1) B_1(-\lambda) + B_{k-m,-\lambda}(2) \right).$$

2. *If* $n = k$, *then we have*

$$B_k(\lambda) = \sum_{m=0}^{k} \binom{k}{n} (1)^{k+m} (1)_{k,\lambda} \left((1)_{k-m,-\lambda}(k-m)(-\lambda - 1) B_1(-\lambda) + B_{k-m,-\lambda}(2) \right). \tag{57}$$

3. Remark

Let us assume that the probability of success in an experiment is p. We wondered if we could say the probability of success in the 9th trial is still p after failing eight times in a ten trial experiment, because there is a psychological burden to be successful. It seems plausible that the probability is less than p. The degenerate Bernstein polynomial $B_n(x|\lambda)$ is used in the probability of success. Thus, we give examples in our results as follows:

Example 1. *Let* $n = 2$, *we have*

$$B_2(2|\lambda) = 2(1)_{1,\lambda}(\lambda - 1) B_1(\lambda) + B_2(\lambda)$$

$$= 2(\lambda - 1)\left(-\frac{1}{2}\right) + \frac{\lambda}{2} + \frac{1}{6}$$

$$= -\frac{\lambda}{2} + \frac{7}{6}.$$

Example 2. *Let* $n = 1$, *we have*

$$B_1(1 - x|\lambda) = (1 - x)_{1,\lambda} + \sum_{m=1}^{1} B_{m,1}(x|\lambda) B_m(\lambda) \frac{(-1)^{1-m}}{(m-1)!} \sum_{k=0}^{m-1} (-1)^k \binom{m-1}{k} \frac{1}{x - k\lambda}$$

$$= (1 - x)_{1,\lambda} + B_{1,1}(x|\lambda) B_1(\lambda) \frac{1}{x}$$

$$= -x + \frac{1}{2}.$$

Example 3. *Let* $n = 1, k = 2$, *we have*

$$(x)_{1,\lambda} \sum_{m=0}^{2} \binom{2}{m} (-1)^{m-2} B_1(1 - x + m|\lambda) = x \left(B_1(1 - x|\lambda) - 2B_1(2 - x|\lambda) + B_1(3 - x|\lambda) \right)$$

$$= -x \left(\left(-x + \frac{1}{2}\right) - 2\left(-x + \frac{3}{2}\right) + \left(-x + \frac{5}{2}\right) \right)$$

$$= 0.$$

4. Conclusions

In this paper, we studied the fully degenerate Bernoulli polynomials associated with degenerate Bernstein polynomials. In Section 1, Equations (12), (18), (20) and (21) are some properties of them. In Section 2, Theorems 1–3 are results of identities for fully degenerate Bernoulli polynomials in connection with degenerate Bernstein polynomials by means of bosonic p-adic integrals on $\mathbb{Z}_p$

and generating functions. Theorems 4–6 are results of higher-order fully Bernoulli polynomials in connection with two variable degenerate Bernstein polynomials by means of bosonic p-adic integrals on $\mathbb{Z}_p$ and generating functions.

Author Contributions: Conceptualization, W.K. and L.-C.J.; Data curation, L.-C.J.; Formal analysis, L.-C.J.; Funding acquisition, J.G.L.; Investigation, J.G.L., W.K. and L.-C.J.; Methodology, W.K. and L.-C.J.; Project administration, L.-C.J.; Resources, L.-C.J.; Supervision, L.-C.J.; Visualization, L.-C.J.; Writing—original draft, W.K. and L.-C.J.; Writing—review & editing, J.G.L. and L.-C.J.

Funding: This paper was supported by Wonkwang University in 2018.

Conflicts of Interest: The authors declare that they have no competing interests.

References

1. Kim, T. Barnes' type multiple degenerate Bernoulli and Euler polynomials. *Appl. Math. Comput.* **2015**, *258*, 556–564. [CrossRef]
2. Kim, D.S.; Kim, T. Identities for degenerate Bernoulli polynomials and Korobov polynomials of the first kind. *Sci. China Math.* **2019**, *62*, 999–1028. [CrossRef]
3. Kim, D.S.; Kim, H.Y.; Kim, D.J.; Kim, T. Identities of Symmetry for Type 2 Bernoulli and Euler Polynomials. *Symmetry* **2019**, *11*, 613. [CrossRef]
4. Kim, D.S.; Kim, T. Degenerate Bernstein polynomials. *Rev. R. Acad. Cienc. Exactas Fís. Nat. Ser. A Mater.* **2018**, 1–8. [CrossRef]
5. Kim, D.S.; Kim, T. Correction to: Degenerate Bernstein polynomials. *Rev. R. Acad. Cienc. Exactas Fís. Nat. Ser. A Mater.* **2019**, 1–2. [CrossRef]
6. Kim, D.S.; Kim, T. Some Identities on Degenerate Bernstein and Degenerate Euler Polynomials. *Mathematics* **2019**, *7*, 47. . [CrossRef]
7. Kim, T. q-Volkenborn integration. *Russ. J. Math. Phys.* **2002**, *9*, 288–299.
8. Kim, D.S.; Kim, T.; Mansour, T.; Seo, J.-J. Fully degenerate poly-Bernoulli polynomials with a q parameter. *Filomat* **2016**, *30*, 1029–1035. [CrossRef]
9. Kim, D.S.; Kim, T. A Note on polyexponential and unipoly functions. *Russ. J. Math. Phys.* **2019**, *26*, 40–49. [CrossRef]
10. Kim, T.; Kim, D.S. Extended Stirling numbers of the first kind associated with Daehee numbers and polynomials. *Rev. R. Acad. Cienc. Exactas Fis. Nat. Ser. A Mater.* **2019**, *113*, 1159–1171. [CrossRef]
11. Pyo, S.-S.; Kim, T.; Rim, S.-H. Degenerate Daehee Numbers of the Third Kind. *Mathematics* **2018**, *6*, 239. [CrossRef]
12. Kim, D.S.; Kim, T.; Seo, J.-J. Fully degenerate poly-Bernoulli numbers and polynomials. *Open Math.* **2016**, *14*, 545–556. [CrossRef]
13. Kim, T.; Kwon, H.I.; Mansour, T.; Rim, S.-H. Symmetric identities for the fully degenerate Bernoulli polynomials and degenerate Euler polynomials under symmetric group of degree n. *Util. Math.* **2017**, *103*, 61–72.
14. Araci, S.; Acikgoz, M. A note on the Frobenius-Euler numbers and polynomials associated with Bernstein polynomials. *Adv. Stud. Contemp. Math. (Kyungshang)* **2012**, *22*, 399–406.
15. Bayad, A.; Kim, T. Identities involving values of Bernstein, q-Bernoulli, and q-Euler polynomials. *Russ. J. Math. Phys.* **2011**, *18*, 133–143. [CrossRef]
16. Choi, J. A note on p-adic integrals associated with Bernstein and q–Bernstein polynomials'. *Adv. Stud. Contemp. Math. (Kyungshang)* **2011**, *21*, 133–138.
17. Kim, D.S.; Kim, T.; Jang, G.W.; Kwon, J. A note on degenerate Bernstein polynomials. *J. Inequal. Appl.* **2019**, *2019*, 129. [CrossRef]
18. Kim, T.; Choi, J.; Kim, Y.H.; Ryoo, C.S. On the fermionic p-adic integral representation of Bernstein polynomials associated with Euler numbers and polynomials. *J. Inequal. Appl.* **2010**, *1*, 864247. [CrossRef]
19. Kim, T.; Choi, J.; Kim, Y.H. On the k-dimensional generalization of q-Bernstein polynomials. *Proc. Jangjeon Math. Soc.* **2011**, *14*, 199–207.
20. Kim, T. A note on q-Bernstein polynomials. *Russ. J. Math. Phys.* **2011**, *18*, 73–82. [CrossRef]

21. Kim, T.; Kim, Y.H.; Bayad, A. A study on the *p*-adic *q*-integral representation on Z p associated with the weighted *q*-Bernstein and *q*-Bernoulli polynomials. *J. Inequal. Appl.* **2011**, 513821. [CrossRef]
22. Kim, T.; Lee, B.; Lee, S.-H.; Rim, S.-H. A note on *q*-Bernstein polynomials associated with *p*-adic integral on $\mathbb{Z}_p$. *J. Comput. Anal. Appl.* **2013**, *15*, 584–592.
23. Kim, W.J.; Kim, D.S.; Kim, H.Y.; Kim, T. Some identities of degenerate Euler polynomials associated with Degenerate Bernstein polynomials. *arXiv* **2019**, arXiv:1904.08592.
24. Kurt, V. Some relation between the Bernstein polynomials and second kind Bernoulli polynomials. *Adv. Stud. Contemp. Math. (Kyungshang)* **2013**, *23*, 43–48.
25. Ostrovska, S. On the *q*–Bernstein polynomials. *Adv. Stud. Contemp. Math. (Kyungshang)* **2005**, *11*, 193–204.
26. Park, J.-W.; Pak, H.K.; Rim, S.-H.; Kim, T.; Lee, S.-H. A note on the *q*-Bernoulli numbers and *q*–Bernstein polynomials. *J. Comput. Anal. Appl.* **2013**, *15*, 722–729.
27. Siddiqui, M.A.; Agrawal, R.R.; Gupta, N. On a class of modified new Bernstein operators. *Adv. Stud. Contemp. Math. (Kyungshang)* **2014**, *24*, 97–107.

Article

The Power Sums Involving Fibonacci Polynomials and Their Applications

Li Chen [1] and Xiao Wang [1,*]

[1]   School of Mathematics, Northwest University, Xi'an 710127, Shaanxi, China; cl1228@stumail.nwu.edu.cn
*   Correspondence: wangxiao_0606@stumail.nwu.edu.cn;

Received: 28 March 2019 ; Accepted: 25 April 2019; Published: 6 May 2019

Abstract: The Girard and Waring formula and mathematical induction are used to study a problem involving the sums of powers of Fibonacci polynomials in this paper, and we give it interesting divisible properties. As an application of our result, we also prove a generalized conclusion proposed by R. S. Melham.

Keywords: Fibonacci polynomials; Lucas polynomials; sums of powers; divisible properties; R. S. Melham's conjectures

MSC: 11B39

---

1. Introduction

For any integer $n \geq 0$, the famous Fibonacci polynomials $\{F_n(x)\}$ and Lucas polynomials $\{L_n(x)\}$ are defined as $F_0(x) = 0$, $F_1(x) = 1$, $L_0(x) = 2$, $L_1(x) = x$ and $F_{n+2}(x) = xF_{n+1}(x) + F_n(x)$, $L_{n+2}(x) = xL_{n+1}(x) + L_n(x)$ for all $n \geq 0$. Now, if we let $\alpha = \frac{x+\sqrt{x^2+4}}{2}$ and $\beta = \frac{x-\sqrt{x^2+4}}{2}$, then it is easy to prove that

$$F_n(x) = \frac{1}{\alpha - \beta}\left(\alpha^n - \beta^n\right) \text{ and } L_n(x) = \alpha^n + \beta^n \text{ for all } n \geq 0.$$

If $x = 1$, we have that $\{F_n(x)\}$ turns into Fibonacci sequences $\{F_n\}$, and $\{L_n(x)\}$ turns into Lucas sequences $\{L_n\}$. If $x = 2$, then $F_n(2) = P_n$, the nth Pell numbers, they are defined by $P_0 = 0$, $P_1 = 1$ and $P_{n+2} = 2P_{n+1} + P_n$ for all $n \geq 0$. In fact, $\{F_n(x)\}$ is a second-order linear recursive polynomial, when x takes a different value x_0, then $F_n(x_0)$ can become a different sequence.

Since the Fibonacci numbers and Lucas numbers occupy significant positions in combinatorial mathematics and elementary number theory, they are thus studied by plenty of researchers, and have gained a large number of vital conclusions; some of them can be found in References [1–15]. For example, Yi Yuan and Zhang Wenpeng [1] studied the properties of the Fibonacci polynomials, and proved some interesting identities involving Fibonacci numbers and Lucas numbers. Ma Rong and Zhang Wenpeng [2] also studied the properties of the Chebyshev polynomials, and obtained some meaningful formulas about the Chebyshev polynomials and Fibonacci numbers. Kiyota Ozeki [3] got some identity involving sums of powers of Fibonacci numbers. That is, he proved that

$$\sum_{k=1}^{n} F_{2k}^{2m+1} = \frac{1}{5^m} \sum_{j=0}^{m} \frac{(-1)^j}{L_{2m+1-2j}} \binom{2m+1}{j} \left(F_{(2m+1-2j)(2n+1)} - F_{2m+1-2j}\right).$$

Helmut Prodinger [4] extended the result of Kiyota Ozeki [3].

In addition, regarding many orthogonal polynomials and famous sequences, Kim et al. have done a lot of important research work, obtaining a series of interesting identities. Interested readers can refer to References [16–22]; we will not list them one by one.

In this paper, our main purpose is to care about the divisibility properties of the Fibonacci polynomials. This idea originated from R. S. Melham. In fact, in [5], R. S. Melham proposed two interesting conjectures as follows:

Conjecture 1. *If $m \geq 1$ is a positive integer, then the summation*

$$L_1 L_3 L_5 \cdots L_{2m+1} \sum_{k=1}^{n} F_{2k}^{2m+1}$$

can be written as $(F_{2n+1} - 1)^2 P_{2m-1} (F_{2n+1})$, where $P_{2m-1}(x)$ is an integer coefficients polynomial with degree $2m - 1$.

Conjecture 2. *If $m \geq 0$ is an integer, then the summation*

$$L_1 L_3 L_5 \cdots L_{2m+1} \sum_{k=1}^{n} L_{2k}^{2m+1}$$

can be written as $(L_{2n+1} - 1) Q_{2m} (L_{2n+1})$, where $Q_{2m}(x)$ is an integer coefficients polynomial with degree $2m$.

Wang Tingting and Zhang Wenpeng [6] solved Conjecture 2 completely. They also proved a weaker conclusion for Conjecture 1. That is,

$$L_1 L_3 L_5 \cdots L_{2m+1} \sum_{k=1}^{n} F_{2k}^{2m+1}$$

can be expressed as $(F_{2n+1} - 1) P_{2m} (F_{2n+1})$, where $P_{2m}(x)$ is a polynomial of degree $2m$ with integer coefficients.

Sun et al. [7] solved Conjecture 1 completely. In fact, Ozeki [3] and Prodinger [4] indicated that the odd power sum of the first several consecutive Fibonacci numbers of even order is equivalent to the polynomial estimated at a Fibonacci number of odd order. Sun et al. in [7] proved that this polynomial and its derivative both disappear at 1, and it can be an integer polynomial when a product of the first consecutive Lucas numbers of odd order multiplies it. This presents an affirmative answer to Conjecture 1 of Melham.

In this paper, we are going to use a new and different method to study this problem, and give a generalized conclusion. That is, we will use the Girard and Waring formula and mathematical induction to prove the conclusions in the following:

Theorem 1. *If n and h are positive integers, then we have the congruence*

$$L_1(x)L_3(x) \cdots L_{2n+1}(x) \sum_{m=1}^{h} F_{2m}^{2n+1}(x) \equiv 0 \bmod (F_{2h+1}(x) - 1)^2 .$$

Taking $x = 1$ and $x = 2$ in Theorem 1, we can instantly infer the two corollaries:

Corollary 1. *Let F_n and L_n be Fibonacci numbers and Lucas numbers, respectively. Then, for any positive integers n and h, we have the congruence*

$$L_1 L_3 L_5 \cdots L_{2n+1} \sum_{m=1}^{h} F_{2m}^{2n+1} \equiv 0 \bmod (F_{2h+1} - 1)^2 .$$

Corollary 2. *Let P_n be nth Pell numbers. Then, for any positive integers n and h, we have the congruence*

$$L_1(2)L_3(2)L_5(2)\cdots L_{2n+1}(2)\sum_{m=1}^{h} P_{2m}^{2n+1} \equiv 0 \bmod (P_{2h+1}-1)^2,$$

where $L_n(2) = \left(1+\sqrt{2}\right)^n + \left(1-\sqrt{2}\right)^n$ is called nth Pell–Lucas numbers.

It is clear that our Corollary 1 gave a new proof for Conjecture 1.

2. Several Lemmas

In this part, we will give four simple lemmas, which are essential to prove our main results.

Lemma 1. *Let h be any positive integer; then, we have*

$$\left(x^2+4, F_{2h+1}(x)-1\right) = 1,$$

where x^2+4 and $F_{2h+1}(x)-1$ are said to be relatively prime.

Proof. From the definition of $F_n(x)$ and binomial theorem, we have

$$
\begin{aligned}
F_{2h+1}(x) &= \frac{1}{2^{2h+1}\sqrt{x^2+4}} \sum_{k=0}^{2h+1} \binom{2h+1}{k} x^k \left(x^2+4\right)^{\frac{2h+1-k}{2}} \\
&\quad - \frac{1}{2^{2h+1}\sqrt{x^2+4}} \sum_{k=0}^{2h+1} \binom{2h+1}{k} x^k (-1)^{2h+1-k} \left(x^2+4\right)^{\frac{2h+1-k}{2}} \\
&= \frac{1}{4^h} \sum_{k=0}^{h} \binom{2h+1}{2k} x^{2k} \left(x^2+4\right)^{h-k}.
\end{aligned}
\tag{1}
$$

Thus, from Equation (1), we have the polynomial congruence

$$4^h F_{2h+1}(x) = \sum_{k=0}^{h} \binom{2h+1}{2k} x^{2k} \left(x^2+4\right)^{h-k} \equiv (2h+1)x^{2h}$$

$$\equiv (2h+1)\left(x^2+4-4\right)^h \equiv (2h+1)(-4)^h \bmod (x^2+4)$$

or

$$F_{2h+1}(x) - 1 \equiv (2h+1)(-1)^h - 1 \bmod (x^2+4).\tag{2}$$

Since x^2+4 is an irreducible polynomial of x, and $(2h+1)(-1)^h - 1$ is not divisible by (x^2+4) for all integer $h \geq 1$, so, from (2), we can deduce that

$$\left(x^2+4, F_{2h+1}(x)-1\right) = 1.$$

Lemma 1 is proved. $\square$

Lemma 2. *Let h and n be non-negative integers with $h \geq 1$; then, we have*

$$(x^2+4)F_{(2h+1)(2n+1)}(x) - L_{2n}(x) - L_{2n+2}(x) \equiv 0 \bmod (F_{2h+1}(x)-1).$$

Proof. We use mathematical induction to calculate the polynomial congruence for n. Noting $L_0(x) = 2$, $L_1(x) = x$, $L_2(x) = x^2 + 2$. Thus, if $n = 0$, then

$$(x^2 + 4)F_{(2h+1)(2n+1)}(x) - L_{2n}(x) - L_{2n+2}(x)$$
$$= (x^2 + 4)F_{2h+1}(x) - 2 - x^2 - 2$$
$$= (x^2 + 4)(F_{2h+1}(x) - 1) \equiv 0 \bmod (F_{2h+1}(x) - 1).$$

If $n = 1$, then $L_2(x) + L_4(x) = x^2 + 2 + x^4 + 4x^2 + 2 = x^4 + 5x^2 + 4$. Note that the identity $F_{2h+1}^3(x) = \frac{1}{x^2+4}\left(F_{3(2h+1)}(x) + 3F_{2h+1}(x)\right)$, so we obtain the congruence

$$(x^2 + 4)F_{(2h+1)(2n+1)}(x) - L_{2n}(x) - L_{2n+2}(x)$$
$$= (x^2 + 4)F_{3(2h+1)}(x) - x^4 - 5x^2 - 4$$
$$= (x^2 + 4)\left[(x^2 + 4)F_{2h+1}^3(x) - 3F_{2h+1}(x)\right] - x^4 - 5x^2 - 4$$
$$= (x^2 + 4)^2\left[F_{2h+1}^3(x) - F_{2h+1}(x)\right] + (x^2 + 4)(x^2 + 1)F_{2h+1}(x) - x^4 - 5x^2 - 4$$
$$\equiv (x^2 + 4)^2\left(F_{2h+1}^2(x) + F_{2h+1}(x)\right)(F_{2h+1}(x) - 1) \equiv 0 \bmod (F_{2h+1}(x) - 1),$$

which means that Lemma 2 is correct for $n = 0$ and 1.

Assume Lemma 2 is right for all integers $n = 0, 1, 2, \cdots, k$. Namely,

$$(x^2 + 4)F_{(2h+1)(2n+1)}(x) - L_{2n}(x) - L_{2n+2}(x) \equiv 0 \bmod (F_{2h+1}(x) - 1), \tag{3}$$

where $0 \le n \le k$.

Thus, $n = k + 1 \ge 2$, and we notice that

$$L_{2(2h+1)}(x)F_{(2h+1)(2k+1)}(x) = F_{(2h+1)(2k+3)}(x) + F_{(2h+1)(2k-1)}(x),$$

$$L_{2k+2}(x) + L_{2k+4}(x) = (x^2 + 2)L_{2k}(x) + (x^2 + 2)L_{2k+2}(x) - (L_{2k-2}(x) + L_{2k}(x))$$

and

$$L_{2(2h+1)}(x) = (x^2 + 4)F_{2h+1}^2(x) - 2 \equiv x^2 + 2 \bmod (F_{2h+1}(x) - 1).$$

From inductive assumption (3), we have

$$(x^2 + 4)F_{(2h+1)(2n+1)}(x) - L_{2n}(x) - L_{2n+2}(x)$$
$$= (x^2 + 4)F_{(2h+1)(2k+3)}(x) - L_{2k+2}(x) - L_{2k+4}(x)$$
$$= (x^2 + 4)L_{2(2h+1)}(x)F_{(2h+1)(2k+1)}(x) - (x^2 + 4)F_{(2h+1)(2k-1)} - L_{2k+2}(x) - L_{2k+4}(x)$$
$$\equiv (x^2 + 4)(x^2 + 2)F_{(2h+1)(2k+1)}(x) - (x^2 + 2)L_{2k}(x) - (x^2 + 2)L_{2k+2}(x)$$
$$\quad - (x^2 + 4)F_{(2h+1)(2k-1)}(x) + L_{2k-2}(x) + L_{2k}(x)$$
$$\equiv (x^2 + 2)\left[(x^2 + 4)F_{(2h+1)(2k+1)}(x) - L_{2k}(x) - L_{2k+2}(x)\right]$$
$$\quad - \left[(x^2 + 4))F_{(2h+1)(2k-1)}(x) - L_{2k-2}(x) - L_{2k}(x)\right]$$
$$\equiv 0 \bmod (F_{2h+1}(x) - 1).$$

Now, we have achieved the results of Lemma 2. $\square$

Lemma 3. *Let h and n be non-negative integers with $h \geq 1$; then, we have the polynomial congruence*

$$L_1(x)L_3(x)\cdots L_{2n+1}(x) \sum_{m=1}^{h} \left[F_{2m(2n+1)}(x) - (2n+1)F_{2m}(x) \right]$$
$$\equiv \; 0 \bmod \left(F_{2h+1}(x) - 1 \right)^2.$$

Proof. For positive integer n, first note that $\alpha\beta = -1$, $L_n(x) = \alpha^n + \beta^n$,

$$\sum_{m=1}^{h} F_{2m(2n+1)}(x) = \frac{1}{\sqrt{x^2+4}} \sum_{m=1}^{h} \left[\alpha^{2m(2n+1)} - \beta^{2m(2n+1)} \right]$$
$$= \frac{1}{\sqrt{x^2+4}} \left[\frac{\alpha^{2(2n+1)}\left(\alpha^{2h(2n+1)} - 1 \right)}{\alpha^{2(2n+1)} - 1} - \frac{\beta^{2(2n+1)}\left(\beta^{2h(2n+1)} - 1 \right)}{\beta^{2(2n+1)} - 1} \right]$$
$$= \frac{1}{L_{2n+1}(x)} \left[F_{(2h+1)(2n+1)}(x) - F_{2n+1}(x) \right] \tag{4}$$

and

$$\sum_{m=1}^{h} F_{2m}(x) = \frac{1}{\sqrt{x^2+4}} \sum_{m=1}^{h} \left[\alpha^{2m} - \beta^{2m} \right] = \frac{1}{L_1(x)} \left[F_{(2h+1)}(x) - 1 \right]. \tag{5}$$

Thus, from Labels (4) and (5), we know that, to prove Lemma 3, now we need to obtain the polynomial congruence

$$L_1(x) \left(F_{(2h+1)(2n+1)}(x) - F_{2n+1}(x) \right) - (2n+1)L_{2n+1}(x)\left(F_{2h+1}(x) - 1 \right)$$
$$\equiv \; 0 \bmod \left(F_{2h+1}(x) - 1 \right)^2. \tag{6}$$

Now, we prove (6) by mathematical induction. If $n = 0$, then it is obvious that (6) is correct. If $n = 1$, we notice that $L_1(x) = x$, $F_{3(2h+1)}(x) = (x^2+4)F_{2h+1}^3(x) - 3F_{2h+1}(x)$ and $F_{2h+1}^3(x) \equiv (F_{2h+1}(x) - 1 + 1)^3 \equiv 3F_{2h+1}(x) - 2 \bmod \left(F_{2h+1}(x) - 1 \right)^2$ we have

$$L_1(x)F_{(2h+1)(2n+1)}(x) - L_1(x)F_{2n+1}(x) - (2n+1)L_{2n+1}(x)\left(F_{2h+1}(x) - 1 \right)$$
$$= xF_{3(2h+1)}(x) - xF_3(x) - 3L_3(x)\left(F_{2h+1}(x) - 1 \right)$$
$$= x(x^2+4)F_{2h+1}^3(x) - 3xF_{2h+1}(x) - x(x^2+1) - 3(x^3+3x)\left(F_{2h+1}(x) - 1 \right)$$
$$\equiv (x^3+4x)\left(3F_{2h+1}(x) - 2 \right) - 3xF_{2h+1}(x) - (x^3+x) - 3(x^3+3x)\left(F_{2h+1}(x) - 1 \right)$$
$$\equiv 3(x^3+3x)\left(F_{2h+1}(x) - 1 \right) - 3(x^3+3x)\left(F_{2h+1}(x) - 1 \right)$$
$$\equiv 0 \bmod \left(F_{2h+1}(x) - 1 \right)^2.$$

Thus, $n = 1$ is fit for (6). Assume that (6) is correct for all integers $n = 0, 1, 2, \cdots, k$. Namely,

$$L_1(x) \left(F_{(2h+1)(2n+1)}(x) - F_{2n+1}(x) \right) - (2n+1)L_{2n+1}(x)\left(F_{2h+1}(x) - 1 \right)$$
$$\equiv \; 0 \bmod \left(F_{2h+1}(x) - 1 \right)^2 \tag{7}$$

for all $n = 0, 1, \cdots, k$.

Where $n = k+1 \geq 2$, we notice

$$L_{2(2h+1)}(x)F_{(2h+1)(2k+1)}(x) = F_{(2h+1)(2k+3)}(x) + F_{(2h+1)(2k-1)}(x)$$

and

$$
\begin{aligned}
L_{2(2h+1)}(x) &= (x^2+4)F_{2h+1}^2(x) - 2 = (x^2+4)\left(F_{2h+1}(x)-1+1\right)^2 - 2 \\
&= (x^2+4)\left[(F_{2h+1}(x)-1)^2 + 2(F_{2h+1}(x)-1)\right] + x^2 + 2 \\
&\equiv 2(x^2+4)(F_{2h+1}(x)-1) + x^2 + 2 \bmod (F_{2h+1}(x)-1)^2.
\end{aligned}
$$

From inductive assumption (7) and Lemma 2, we have

$$
\begin{aligned}
& xF_{(2h+1)(2n+1)}(x) - xF_{2n+1}(x) - (2n+1)L_{2n+1}(x)\left(F_{2h+1}(x)-1\right) \\
=\ & xF_{(2h+1)(2k+3)}(x) - xF_{2k+3}(x) - (2k+3)L_{2k+3}(x)\left(F_{2h+1}(x)-1\right) \\
=\ & xL_{2(2h+1)}(x)F_{(2h+1)(2k+1)}(x) - xF_{(2h+1)(2k-1)}(x) - xF_{2k+3}(x) \\
& -(2k+3)L_{2k+3}(x)\left(F_{2h+1}(x)-1\right) \\
\equiv\ & 2x(x^2+4)(F_{2h+1}(x)-1)F_{(2h+1)(2k+1)}(x) + x(x^2+2)F_{(2h+1)(2k+1)}(x) \\
& -xF_{(2h+1)(2k-1)}(x) - x(x^2+2)F_{2k+1}(x) + xF_{2k-1}(x) \\
& -(x^2+2)(2k+1)L_{2k+1}(x)\left(F_{2h+1}(x)-1\right) + (2k-1)L_{2k-1}(x)\left(F_{2h+1}(x)-1\right) \\
& -2x\left(L_{2k}(x) + L_{2k+2}(x)\right)\left(F_{2h+1}(x)-1\right) \\
\equiv\ & 2x(F_{2h+1}(x)-1)\left[(x^2+4)F_{(2h+1)(2k+1)}(x) - L_{2k}(x) - L_{2k+2}(x)\right] \\
& +(x^2+2)\left[xF_{(2h+1)(2k+1)}(x) - xF_{2k+1}(x) - (2k+1)L_{2k+1}(x)\left(F_{2h+1}(x)-1\right)\right] \\
& -\left[xF_{(2h+1)(2k-1)}(x) - xF_{2k-1}(x) - (2k-1)L_{2k-1}(x)\left(F_{2h+1}(x)-1\right)\right] \\
\equiv\ & 2x(F_{2h+1}(x)-1)\left[(x^2+4)F_{(2h+1)(2k+1)}(x) - L_{2k}(x) - L_{2k+2}(x)\right] \\
\equiv\ & 0 \bmod (F_{2h+1}(x)-1)^2.
\end{aligned}
$$

Now, we attain Lemma 3 by mathematical induction. $\square$

Lemma 4. *For all non-negative integers u and real numbers X, Y, we have the identity*

$$
X^u + Y^u = \sum_{k=0}^{\left[\frac{u}{2}\right]} (-1)^k \frac{u}{u-k}\binom{u-k}{k}(X+Y)^{u-2k}(XY)^k,
$$

in which $[x]$ denotes the greatest integer $\leq x$.

Proof. This formula due to Waring [15]. It can also be found in Girard [14]. $\square$

3. Proof of the Theorem

We will achieve the theorem by these lemmas. Taking $X = \alpha^{2m}$, $Y = -\beta^{2m}$ and $U = 2n+1$ in Lemma 4, we notice that $XY = -1$, from the expression of $F_n(x)$

$$
\begin{aligned}
F_{2m(2n+1)}(x) &= \sum_{k=0}^{n}(-1)^k \frac{2n+1}{2n+1-k}\binom{2n+1-k}{k}(x^2+4)^{n-k}F_{2m}^{2n+1-2k}(x)(-1)^k \\
&= \sum_{k=0}^{n} \frac{2n+1}{2n+1-k}\binom{2n+1-k}{k}(x^2+4)^{n-k}F_{2m}^{2n+1-2k}(x).
\end{aligned} \tag{8}
$$

For any integer $h \geq 1$, from (8), we get

$$\sum_{m=1}^{h} \left[F_{2m(2n+1)}(x) - (2n+1)F_{2m}(x) \right]$$

$$= \sum_{k=0}^{n-1} \frac{2n+1}{2n+1-k} \binom{2n+1-k}{k} (x^2+4)^{n-k} \sum_{m=1}^{h} F_{2m}^{2n+1-2k}(x). \tag{9}$$

If $n = 1$, then, from (9), we can get

$$L_1(x)L_3(x) \sum_{m=1}^{h} (F_{6m}(x) - 3F_{2m}(x)) = L_1(x)L_3(x)(x^2+4) \sum_{m=1}^{h} F_{2m}^3(x). \tag{10}$$

From Lemma 1, we know that $(x^2+4, F_{2h+1}(x) - 1) = 1$, so, applying Lemma 3 and (10), we deduce that

$$L_1(x)L_3(x) \sum_{m=1}^{h} F_{2m}^3(x) \equiv 0 \bmod (F_{2h+1}(x) - 1)^2. \tag{11}$$

This means that Theorem 1 is suitable for $n = 1$.

Assume that Theorem 1 is correct for all integers $n = 1, 2, \cdots, s$. Then,

$$L_1(x)L_3(x) \cdots L_{2n+1}(x) \sum_{m=1}^{h} F_{2m}^{2n+1}(x) \equiv 0 \bmod (F_{2h+1}(x) - 1)^2 \tag{12}$$

for all integers $1 \leq n \leq s$.

When $n = s + 1$, from (9), we obtain

$$\sum_{m=1}^{h} \left(F_{2m(2s+3)}(x) - (2s+3)F_{2m}(x) \right)$$

$$= \sum_{k=0}^{s} \frac{2s+3}{2s+3-k} \binom{2s+3-k}{k} (x^2+4)^{s+1-k} \sum_{m=1}^{h} F_{2m}^{2s+3-2k}(x)$$

$$= \sum_{k=1}^{s} \frac{2s+3}{2s+3-k} \binom{2s+3-k}{k} (x^2+4)^{s+1-k} \sum_{m=1}^{h} F_{2m}^{2s+3-2k}(x)$$

$$+ (x^2+4)^{s+1} \sum_{m=1}^{h} F_{2m}^{2s+3}(x). \tag{13}$$

From Lemma 3, we have

$$L_1(x)L_3(x) \cdots L_{2s+3}(x) \sum_{m=1}^{h} \left[F_{2m(2s+3)}(x) - (2s+3)F_{2m}(x) \right]$$

$$\equiv 0 \bmod (F_{2h+1}(x) - 1)^2. \tag{14}$$

Applying inductive hypothesis (12), we obtain

$$L_1(x)L_3(x) \cdots L_{2s+1}(x) \sum_{k=1}^{s} \frac{2s+3}{2s+3-k} \binom{2s+3-k}{k}$$

$$\times (x^2+4)^{s+1-k} \sum_{m=1}^{h} F_{2m}^{2s+3-2k}(x) \equiv 0 \bmod (F_{2h+1}(x) - 1)^2. \tag{15}$$

Combining (13), (14), (15) and Lemma 3, we have the conclusion

$$L_1(x)L_3(x)\cdots L_{2s+3}(x) \cdot (x^2+4)^{s+1} \sum_{m=1}^{h} F_{2m}^{2s+3}(x)$$

$$\equiv 0 \bmod (F_{2h+1}(x)-1)^2. \tag{16}$$

Note that $\left(x^2+4, F_{2h+1}(x)-1\right) = 1$, so (16) indicates the conclusion

$$L_1(x)L_3(x)\cdots L_{2s+3}(x) \cdot \sum_{m=1}^{h} F_{2m}^{2s+3}(x) \equiv 0 \bmod (F_{2h+1}(x)-1)^2.$$

Now, we apply mathematical induction to achieve Theorem 1.

Author Contributions: Conceptualization, L.C.; methodology, L.C and X.W.; validation, L.C. and X.W.; formal analysis, L.C.; investigation, X.W.; resources, L.C.; writing—original draft preparation, L.C.; writing—review and editing, X.W.; visualization,L.C.; supervision, L.C.; project administration, X.W.; all authors have read and approved the final manuscript.

Funding: This work is supported by the N. S. F. (11771351) and (11826205) of P. R. China.

Acknowledgments: The authors would like to thank the referees for their very helpful and detailed comments, which have significantly improved the presentation of this paper.

Conflicts of Interest: The authors state that there are no conflicts of interest concerning the publication of this paper.

References

1. Yi, Y.; Zhang, W.P. Some identities involving the Fibonacci polynomials. *Fibonacci Q.* **2002**, *40*, 314–318.
2. Ma, R.; Zhang, W.P. Several identities involving the Fibonacci numbers and Lucas numbers. *Fibonacci Q.* **2007**, *45*, 164–170.
3. Ozeki, K. On Melham's sum. *Fibonacci Q.* **2008**, *46*, 107–110.
4. Prodinger, H. On a sum of Melham and its variants. *Fibonacci Q.* **2008**, *46*, 207–215.
5. Melham, R.S. Some conjectures concerning sums of odd powers of Fibonacci and Lucas numbers. *Fibonacci Q.* **2009** , *4*, 312–315.
6. Wang, T.T.; Zhang, W.P. Some identities involving Fibonacci, Lucas polynomials and their applications. *Bull. Math. Soc. Sci. Math. Roumanie* **2012**, *55*, 95–103.
7. Sun, B.Y.; Xie, M.H.Y.; Yang, A.L.B. Melham's Conjecture on Odd Power Sums of Fibonacci Numbers. *Quaest. Math.* **2016**, *39*, 945–957. [CrossRef]
8. Duncan, R.I. Application of uniform distribution to the Fibonacci numbers. *Fibonacci Q.* **1967** , *5*, 137–140.
9. Kuipers, L. Remark on a paper by R. L. Duncan concerning the uniform distribution mod 1 of the sequence of the logarithms of the Fibonacci numbers. *Fibonacci Q.* **1969** , *7*, 465–466.
10. Chen, L.; Zhang, W.P. Chebyshev polynomials and their some interesting applications. *Adv. Differ. Equ.* **2017**, *2017*, 303.
11. Li, X.X. Some identities involving Chebyshev polynomials, Mathematical Problems in Engineering. *Math. Probl. Eng.* **2015**, *2015*. [CrossRef]
12. Ma, Y.K; Lv, X.X. Several identities involving the reciprocal sums of Chebyshev polynomials. *Math. Probl. Eng.* **2017**, *2017*. [CrossRef]
13. Clemente, C. Identities and generating functions on Chebyshev polynomials *Georgian Math. J.* **2012**, *39*, 427–440.
14. Gould, H.W. The Girard-Waring power sum formulas for symmetric functions and Fibonacci sequences. *Fibonacci Q.* **1999**, *37*, 135–140.
15. Waring, E. *Miscellanea Analytica: De Aequationibus Algebraicis, Et Curvarum Proprietatibus*; Nabu Press: Charleston, SC, USA, 2010.
16. Kim, D.S.; Dolgy, D.V.; Kim, D.; Kim, T. Representing by Orthogonal Polynomials for Sums of Finite Products of Fubini Polynomials. *Mathematics* **2019**, *7*, 319. [CrossRef]

17. Kim, T.; Hwang, K.-W.; Kim, D.S.; Dolgy, D.V. Connection Problem for Sums of Finite Products of Legendre and Laguerre Polynomials. *Symmetry* **2019**, *11*, 317. [CrossRef]

18. Kizilates, C.; Cekim, B.; Tuglu, N.; Kim, T. New Families of Three-Variable Polynomials Coupled with Well-Known Polynomials and Numbers. *Symmetry* **2019**, *11*, 264. [CrossRef]

19. Kim, T.; Kim, D.S.; Dolgy, D.V.; Park, J.-W. Sums of finite products of Chebyshev polynomials of the second kind and of Fibonacci polynomials. *J. Inequal. Appl.* **2018**, *2018*, 148. [CrossRef]

20. Kim, T.; Dolgy, D.V.; Kim, D.S.; Seo, J.J. Convolved Fibonacci numbers and their applications. *Ars Combin.* **2017**, *135*, 119–131.

21. Kim, T.; Kim, D.S.; Dolgy, D.V.; Kwon, J. Representing Sums of Finite Products of Chebyshev Polynomials of the First Kind and Lucas Polynomials by Chebyshev Polynomials. *Mathematics* **2019**, *7*, 26. [CrossRef]

22. Kim, T.; Kim, D.S.; Kwon, J.; Dolgy, D.V. Expressing Sums of Finite Products of Chebyshev Polynomials of the Second Kind and of Fibonacci Polynomials by Several Orthogonal Polynomials. *Mathematics* **2018**, *6*, 210. [CrossRef]

 *symmetry*

Article

Identities of Symmetry for Type 2 Bernoulli and Euler Polynomials

Dae San Kim [1], Han Young Kim [2], Dojin Kim [3,*] and Taekyun Kim [2]

[1]   Department of Mathematics, Sogang University, Seoul 04107, Korea; dskim@sogang.ac.kr
[2]   Department of Mathematics, Kwangwoon University, Seoul 01897, Korea; gksdud213@kw.ac.kr (H.Y.K.);
     tkkim@kw.ac.kr (T.K.)
[3]   Department of Mathematics, Pusan National University, Busan 46241, Korea
*   Correspondence: kimdojin@pusan.ac.kr

Received: 13 April 2019; Accepted: 28 April 2019; Published: 2 May 2019

Abstract: The main purpose of this paper is to give several identities of symmetry for type 2 Bernoulli and Euler polynomials by considering certain quotients of bosonic p-adic and fermionic p-adic integrals on $\mathbb{Z}_p$, where p is an odd prime number. Indeed, they are symmetric identities involving type 2 Bernoulli polynomials and power sums of consecutive odd positive integers, and the ones involving type 2 Euler polynomials and alternating power sums of odd positive integers. Furthermore, we consider two random variables created from random variables having Laplace distributions and show their moments are given in terms of the type 2 Bernoulli and Euler numbers.

Keywords: type 2 Bernoulli polynomials; type 2 Euler polynomials; identities of symmetry; Laplace distribution

---

1. Introduction

In this section, we are going to review some known results. We first recall the definitions of Bernoulli and Euler polynomials together with their type 2 polynomials. Then, we introduce the bosonic p-adic integrals and the fermionic p-adic integrals on $\mathbb{Z}_p$ that we need for the derivation of an identity of symmetry. As is well known, the Bernoulli polynomials are defined by

$$\frac{t}{e^t - 1} e^{xt} = \sum_{n=0}^{\infty} B_n(x) \frac{t^n}{n!}, \tag{1}$$

(see [1,2]).

In particular, the Bernoulli numbers are the constant terms $B_n = B_n(0)$ of the Bernoulli polynomials. By making use of (1), we can deduce that

$$\sum_{l=0}^{n-1} l^k = \frac{1}{k+1} (B_{k+1}(n) - B_{k+1}), \text{ for } k = 0, 1, 2, \cdots. \tag{2}$$

The type 2 Bernoulli polynomials are defined by generating function

$$\frac{t}{e^t - e^{-t}} e^{xt} = \sum_{n=0}^{\infty} b_n(x) \frac{t^n}{n!}, \tag{3}$$

(see [3,4]).

In particular, $b_n = b_n(0)$ are called type 2 Bernoulli numbers. From (3), it can be seen that

$$b_n(x) = \sum_{k=0}^{n} \binom{n}{k} b_k x^{n-k},$$

(4)

(see [3,4]).

Analogously to (2), we observe that

$$\sum_{l=0}^{n-1} e^{(2l+1)t} = \frac{1}{e^t - e^{-t}} (e^{2nt} - 1)$$

$$= \sum_{k=0}^{\infty} \left(\frac{b_{k+1}(2n) - b_{k+1}}{k+1} \right) \frac{t^k}{k!}.$$

(5)

Thus, by (5), we get

$$\sum_{l=0}^{n-1} (2l+1)^k = \frac{1}{k+1} (b_{k+1}(2n) - b_{k+1}), \quad k = 0, 1, 2, \cdots.$$

(6)

Let p be a fixed odd prime number. Throughout this paper, we will use the notations $\mathbb{Z}_p, \mathbb{Q}_p, \mathbb{C}_p$, and $\mathbb{C}$ to denote the ring of p-adic rational integers, the field of p-adic rational numbers, the completion of an algebraic closure of $\mathbb{Q}_p$, and the field of complex numbers, respectively. The normalized valuation in $\mathbb{C}_p$ is denoted by $|\cdot|_p$, with $|p|_p = \frac{1}{p}$. For a uniformly differentiable function f on $\mathbb{Z}_p$, the bosonic p-adic integral on $\mathbb{Z}_p$ (or p-adic invariant integral on $\mathbb{Z}_p$) is defined by

$$\int_{\mathbb{Z}_p} f(x) d\mu_0(x) = \lim_{N \to \infty} \sum_{x=0}^{p^N-1} f(x) \mu_0(x + p^N \mathbb{Z}_p) = \lim_{N \to \infty} \frac{1}{p^N} \sum_{x=0}^{p^N-1} f(x).$$

(7)

Then, by (7), we easily get

$$\int_{\mathbb{Z}_p} f(x+1) d\mu_0(x) - \int_{\mathbb{Z}_p} f(x) d\mu_0(x) = f'(0),$$

(8)

(see [5,6]).

The fermionic integral on $\mathbb{Z}_p$ is defined by Kim [6] as

$$\int_{\mathbb{Z}_p} f(x) d\mu_{-1}(x) = \lim_{N \to \infty} \sum_{x=0}^{p^N-1} f(x) \mu_{-1}(x + p^N \mathbb{Z}_p) = \lim_{N \to \infty} \sum_{x=0}^{p^N-1} f(x)(-1)^x.$$

(9)

From (9), we can show that

$$\int_{\mathbb{Z}_p} f(x+1) d\mu_{-1}(x) + \int_{\mathbb{Z}_p} f(x) d\mu_{-1}(x) = 2f(0),$$

(10)

(see [4,7–10]).

It is well known that the Euler polynomials are defined by

$$\frac{2}{e^t + 1} e^{xt} = \sum_{n=0}^{\infty} E_n^*(x) \frac{t^n}{n!}.$$

(11)

We denote the Euler numbers by $E_n^* = E_n^*(0)$, $(n \geq 0)$. Clearly, we have

$$2 \sum_{l=0}^{n-1} (-1)^l e^{lt} = \frac{2}{e^t+1}(e^{nt}+1), \quad \text{where } n \equiv 1 \pmod{2}. \tag{12}$$

From (11) and (12), we obtain that

$$2 \sum_{l=0}^{n-1} (-1)^l l^k = E_k^*(n) + E_k^*, \tag{13}$$

where n is a positive odd integer.

Now, we consider the type 2 Euler polynomials which are given by

$$\frac{2}{e^t+e^{-t}}e^{xt} = \text{sech}(t)e^{xt} = \sum_{n=0}^{\infty} E_n(x)\frac{t^n}{n!}. \tag{14}$$

In particular, when $x = 0$, $E_n = E_n(0)$ are called the type 2 Euler numbers.

In this paper, we obtain some identities of symmetry involving the type 2 Bernoulli polynomials, the type 2 Euler polynomials, power sums of odd positive integers and alternating power sums of odd positive integers which are derived from certain quotients of bosonic p-adic and fermionic p-adic integrals on $\mathbb{Z}_p$. In the following section, we will construct two random variables from random variables having Laplace distributions whose moments are closely related to the type 2 Bernoulli and Euler numbers. All the results in Sections 2 and 3 are newly developed. Finally, we note that the results here have applications in such diverse areas as combinatorics, probability, algebra and analysis (see [11–13]).

2. Some Identities of Symmetry for Type 2 Bernoulli and Euler Polynomials

In virtue of (8), we readily see that

$$\frac{1}{2} \int_{\mathbb{Z}_p} e^{(2x+1)t} d\mu_0(x) = \frac{t}{e^t-e^{-t}}. \tag{15}$$

Hence, by (15), we get

$$\frac{1}{2} \int_{\mathbb{Z}_p} (2x+1)^n d\mu_0(x) = b_n, \quad (n \geq 0). \tag{16}$$

In addition, it follows from (15) that

$$\frac{1}{2} \int_{\mathbb{Z}_p} e^{(2y+x+1)t} d\mu_0(y) = \frac{t}{e^t-e^{-t}}e^{xt} = \sum_{n=0}^{\infty} b_n(x)\frac{t^n}{n!}. \tag{17}$$

Hence, by (17), we get

$$\frac{1}{2} \int_{\mathbb{Z}_p} (2y+x+1)^n d\mu_0(y) = b_n(x), \quad (n \geq 0). \tag{18}$$

Using (15) and (17), one can easily check that

$$\frac{1}{2} \left(\int_{\mathbb{Z}_p} e^{(2x+2n+1)t} d\mu_0(x) - \int_{\mathbb{Z}_p} e^{(2x+1)t} d\mu_0(x) \right) = t \sum_{l=0}^{n-1} e^{(2l+1)t}. \tag{19}$$

Next, we let $T_k(n) = \sum_{l=0}^{n} (2l+1)^k$, $(k \in \mathbb{N} \cup \{0\})$. Note that $T_k(n)$ represents the kth power sums of consecutive positive odd integers. By (19), we easily get

$$\int_{\mathbb{Z}_p} e^{(2x+1+2n)t} d\mu_0(x) - \int_{\mathbb{Z}_p} e^{(2x+1)t} d\mu_0(x) = \frac{2nt \int_{\mathbb{Z}_p} e^{(2x+1)t} d\mu_0(x)}{\int_{\mathbb{Z}_p} e^{2nxt} d\mu_0(x)}. \tag{20}$$

Let w_1, w_2 be positive integers. Then, we observe that

$$\begin{aligned}
\frac{w_1 \int_{\mathbb{Z}_p} e^{(2x+1)t} d\mu_0(x)}{\int_{\mathbb{Z}_p} e^{2w_1 xt} d\mu_0(x)} &= \sum_{l=0}^{w_1-1} e^{(2l+1)t} \\
&= \sum_{k-0}^{\infty} \sum_{l=0}^{w_1-1} (2l+1)^k \frac{t^k}{k!} \tag{21} \\
&= \sum_{k=0}^{\infty} T_k(w_1 - 1) \frac{t^k}{k!}.
\end{aligned}$$

Now, we consider the next quotient of bosonic p-adic integrals on $\mathbb{Z}_p$ from which the identities of symmetry for the type 2 Bernoulli polynomials follow:

$$I(w_1, w_2) = \frac{w_1 w_2}{2} \frac{\int_{\mathbb{Z}_p} \int_{\mathbb{Z}_p} e^{(2w_1 x_1 + w_1 + 2w_2 x_2 + w_2 + w_1 w_2 x)t} d\mu_0(x_1) d\mu_0(x_2)}{\int_{\mathbb{Z}_p} e^{2w_1 w_2 xt} d\mu_0(x)}. \tag{22}$$

From (22), we have

$$\begin{aligned}
I(w_1, w_2) &= \frac{w_2}{2} \int_{\mathbb{Z}_p} e^{(2x_1 + w_2 + 1)w_1 t} d\mu_0(x) \frac{w_1 \int_{\mathbb{Z}_p} e^{(2w_2 x_2 + w_2)t} d\mu_0(x_2)}{\int_{\mathbb{Z}_p} e^{2w_1 w_2 xt} d\mu_0(x)} \\
&= w_2 \sum_{k=0}^{\infty} b_k(w_2 x) \frac{w_1^k}{k!} t^k \sum_{l=0}^{\infty} T_l(w_1 - 1) \frac{w_2^l}{l!} t^l \tag{23} \\
&= \sum_{n=0}^{\infty} \sum_{k=0}^{n} \binom{n}{k} b_k(w_2 x) T_{n-k}(w_1 - 1) w_1^k w_2^{n-k+1} \frac{t^n}{n!}.
\end{aligned}$$

We note from (22) that $I(w_1, w_2) = I(w_2, w_1)$. Interchanging w_1 and w_2, we get

$$I(w_2, w_1) = \sum_{n=0}^{\infty} \sum_{k=0}^{n} \binom{n}{k} b_k(w_1 x) T_{n-k}(w_2 - 1) w_2^k w_1^{n-k+1} \frac{t^n}{n!}. \tag{24}$$

Therefore, by (23) and (24), we obtain the following theorem.

Theorem 1. *For* $w_1, w_2 \in \mathbb{N}$ *and* $n \in \mathbb{N} \cup \{0\}$, *we have*

$$\sum_{k=0}^{n} \binom{n}{k} b_k(w_2 x) T_{n-k}(w_1 - 1) w_1^k w_2^{n-k+1} = \sum_{k=0}^{n} \binom{n}{k} b_k(w_1 x) T_{n-k}(w_2 - 1) w_2^k w_1^{n-k+1}.$$

Setting $x = 0$ in Theorem 1, we obtain the following corollary.

Corollary 1. *For $w_1, w_2 \in \mathbb{N}$ and $n \in \mathbb{N} \cup \{0\}$, we have*

$$\sum_{k=0}^{n} \binom{n}{k} b_k T_{n-k}(w_1 - 1) w_1^k w_2^{n-k+1} = \sum_{k=0}^{n} \binom{n}{k} b_k T_{n-k}(w_2 - 1) w_2^k w_1^{n-k+1}.$$

Furthermore, let us take $w_2 = 1$ in Corollary 1. Then, we have

$$\sum_{k=0}^{n} \binom{n}{k} b_k w_1^{n-k+1} = \sum_{k=0}^{n} \binom{n}{k} b_k T_{n-k}(w_1 - 1) w_1^k. \tag{25}$$

Therefore, by (4) and (25), we obtain the following corollary.

Corollary 2. *For $w_1 \in \mathbb{N}$ and $n \in \mathbb{N} \cup \{0\}$, we have*

$$b_n(w_1) = \sum_{k=0}^{n} \binom{n}{k} b_k T_{n-k}(w_1 - 1) w_1^{k-1} = \sum_{k=0}^{n} \binom{n}{k} b_k w_1^{k-1} \sum_{l=0}^{w_1-1} (2l+1)^{n-k}.$$

From (22), we observe that

$$
\begin{aligned}
I(w_1, w_2) &= \frac{w_2}{2} e^{w_1 w_2 x t} \int_{\mathbb{Z}_p} e^{2w_1 x_1 t + w_1 t} d\mu_0(x_1) \frac{w_1 \int_{\mathbb{Z}_p} e^{(2w_2 x_2 + w_2)t} d\mu_0(x_1)}{\int_{\mathbb{Z}_p} e^{2w_1 w_2 x t} d\mu_0(x)} \\
&= \frac{w_2}{2} e^{w_1 w_2 x t} \int_{\mathbb{Z}_p} e^{(2w_1 x_1 + w_1)t} d\mu_0(x_1) \sum_{l=0}^{w_1-1} e^{(2l+1)w_2 t} \\
&= \frac{w_2}{2} \sum_{l=0}^{w_1-1} \int_{\mathbb{Z}_p} e^{\left(2x_1 + 1 + w_2 x + (2l+1)\frac{w_2}{w_1}\right) w_1 t} d\mu_0(x_1) \\
&= \sum_{n=0}^{\infty} w_2 \sum_{l=0}^{w_1-1} b_n \left(w_2 x + (2l+1)\frac{w_2}{w_1} \right) \frac{w_1^n t^n}{n!}.
\end{aligned}
\tag{26}
$$

By interchanging w_1 and w_2, we obtain the following equation:

$$I(w_2, w_1) = \sum_{n=0}^{\infty} w_1 \sum_{l=0}^{w_2-1} b_n \left(w_1 x + (2l+1)\frac{w_1}{w_2} \right) \frac{w_2^n t^n}{n!}. \tag{27}$$

As $I(w_1, w_2) = I(w_2, w_1)$, the following theorem is immediate from (26) and (27).

Theorem 2. *For $w_1, w_2 \in \mathbb{N}$ and $n \in \mathbb{N} \cup \{0\}$, we have*

$$w_1^n w_2 \sum_{l=0}^{w_1-1} b_n \left(w_2 x + (2l+1)\frac{w_2}{w_1} \right) = w_2^n w_1 \sum_{l=0}^{w_2-1} b_n \left(w_1 x + (2l+1)\frac{w_1}{w_2} \right).$$

Example 1. *We check the result in Theorem 2 in the case of $n = 2$, $w_1 = 3$, and $w_2 = 7$. We first note that $b_2(x) = \frac{1}{2}(x^2 - \frac{1}{3})$. This can be obtained from $B_2(x) = x^2 - x + \frac{1}{6}$ and the relation $b_n(x) = 2^{n-1} B_n(\frac{x+1}{2})$ which follows from (1) and (3). Thus, we have to see that*

$$\sum_{l=0}^{2} \left\{ \left(7x + \frac{7}{3}(2l+1)\right)^2 - \frac{1}{3} \right\} = \frac{7}{3} \sum_{l=0}^{6} \left\{ \left(3x + \frac{3}{7}(2l+1)\right)^2 - \frac{1}{3} \right\}. \tag{28}$$

Now, we can easily show that both the left and the right side of (28) *are equal to* $147x^2 + 294x + \frac{1706}{9}$.

Let us take $w_1 = 1$. Then, by Theorem 2, we get

$$w_2 b_n(w_2 x + w_2) = w_2^n \sum_{l=0}^{w_2-1} b_n\left(x + (2l+1)\frac{1}{w_2}\right). \tag{29}$$

Equivalently, by (29), we have

$$b_n(w_2 x + w_2) = w_2^{n-1} \sum_{l=0}^{w_2-1} b_n\left(x + (2l+1)\frac{1}{w_2}\right). \tag{30}$$

Similarly to (13), we observe that

$$2\sum_{l=0}^{n-1}(-1)^l e^{(2l+1)t} = \sum_{k=0}^{\infty}(E_k + E_k(2n))\frac{t^k}{k!}, \tag{31}$$

where $n \in \mathbb{N}$ with $n \equiv 1 \pmod 2$. Thus, by (31), we get

$$2\sum_{l=0}^{n-1}(-1)^l(2l+1)^k = E_k + E_K(2n), \tag{32}$$

where $k \in \mathbb{N} \cup \{0\}$ and $n \in \mathbb{N}$ with $n \equiv 1 \pmod 2$.

From (14), we easily note that

$$E_n(x) = \sum_{k=0}^{n}\binom{n}{k}E_k x^{n-k}, \quad (n \geq 0). \tag{33}$$

By (10), we get

$$\int_{\mathbb{Z}_p} e^{(2y+x+1)t}d\mu_{-1}(y) = \frac{2}{e^t + e^{-t}}e^{xt} = \sum_{n=0}^{\infty}E_n(x)\frac{t^n}{n!}. \tag{34}$$

Thus, we have

$$\int_{\mathbb{Z}_p}(2y + x + 1)^n d\mu_{-1}(y) = E_n(x), \quad (n \geq 0).$$

The next equation follows immediately from (10):

$$\int_{\mathbb{Z}_p} e^{(2y+2n+1)t}d\mu_{-1}(y) + \int_{\mathbb{Z}_p} e^{(2x+1)t}d\mu_{-1}(x) = 2\sum_{l=0}^{n-1}e^{(2l+1)t}(-1)^l, \tag{35}$$

where $n \in \mathbb{N}$ with $n \equiv 1 \pmod 2$.

Now, we let $A_k(n) = \sum_{l=0}^{n}(-1)^l(2l+1)^k$, $(k \in \mathbb{N} \cup \{0\})$. Here we note that $A_k(n)$ is the alternating kth power sums of consecutive odd positive integers. From (35), we have

$$\int_{\mathbb{Z}_p} e^{(2x+2n+1)t}d\mu_{-1}(x) + \int_{\mathbb{Z}_p} e^{(2x+1)t}d\mu_{-1}(x) = \frac{2\int_{\mathbb{Z}_p} e^{(2x+1)t}d\mu_{-1}(x)}{\int_{\mathbb{Z}_p} e^{2nxt}d\mu_{-1}(x)}. \tag{36}$$

Let a, b be positive integers with $a \equiv 1 \pmod 2$ and $b \equiv 1 \pmod 2$. Then, by using the fermionic p-adic integral on $\mathbb{Z}_p$, we get

$$
\begin{aligned}
\frac{2 \int_{\mathbb{Z}_p} e^{(2x+1)t} d\mu_{-1}(x)}{\int_{\mathbb{Z}_p} e^{2axt} d\mu_{-1}(x)} &= 2 \sum_{l=0}^{a-1} e^{(2l+1)t}(-1)^l \\
&= \sum_{k=0}^{\infty} 2 \sum_{l=0}^{a-1} (2l+1)^k (-1)^l \frac{t^k}{k!} \\
&= \sum_{k=0}^{\infty} 2 A_k(a-1) \frac{t^k}{k!}.
\end{aligned}
\tag{37}
$$

We now consider the next quotient of the fermionic p-adic integrals on $\mathbb{Z}_p$ from which the identities of symmetry for the type 2 Euler polynomials follow:

$$
J(a,b) = \frac{\int_{\mathbb{Z}_p} \int_{\mathbb{Z}_p} e^{(2ax_1+a+2bx_2+b+abx)t} d\mu_{-1}(x_1) d\mu_{-1}(x_2)}{\int_{\mathbb{Z}_p} e^{2abxt} d\mu_{-1}(x)}.
\tag{38}
$$

From (38), we can derive the following equation given by

$$
\begin{aligned}
J(a,b) &= \frac{1}{2} \int_{\mathbb{Z}_p} e^{a(2x_1+1+bx)t} d\mu_{-1}(x_1) \frac{2 \int_{\mathbb{Z}_p} e^{(2bx_2+b)t} d\mu_{-1}(x_2)}{\int_{\mathbb{Z}_p} e^{2abxt} d\mu_{-1}(x)} \\
&= \frac{1}{2} \sum_{k=0}^{\infty} E_k(bx) \frac{a^k t^k}{k!} 2 \sum_{l=0}^{\infty} A_l(a-1) \frac{b^l t^l}{l!} \\
&= \sum_{n=0}^{\infty} \sum_{k=0}^{n} \binom{n}{k} E_k(bx) A_{n-k}(a-1) a^k b^{n-k} \frac{t^n}{n!}.
\end{aligned}
\tag{39}
$$

We note from (38) that $J(a,b) = J(b,a)$. Interchanging a and b, we get

$$
J(b,a) = \sum_{n=0}^{\infty} \sum_{k=0}^{n} \binom{n}{k} E_k(ax) A_{n-k}(b-1) b^k a^{n-k} \frac{t^n}{n!}.
\tag{40}
$$

The following theorem is an immediate consequence of (39) and (40).

Theorem 3. *For $n \geq 0$, $a, b \in \mathbb{N}$ with $a \equiv 1 \pmod 2$ and $b \equiv 1 \pmod 2$, we have*

$$
\sum_{k=0}^{n} \binom{n}{k} E_k(bx) A_{n-k}(a-1) a^k b^{n-k} = \sum_{k=0}^{n} \binom{n}{k} E_k(ax) A_{n-k}(b-1) b^k a^{n-k}.
$$

The next corollary is now obtained by setting $x = 0$ in Theorem 3.

Corollary 3. *For $n \geq 0$, $a, b \in \mathbb{N}$, with $a \equiv 1 \pmod 2$ and $b \equiv 1 \pmod 2$, we have*

$$
\sum_{k=0}^{n} \binom{n}{k} E_k A_{n-k}(a-1) a^k b^{n-k} = \sum_{k=0}^{n} \binom{n}{k} E_k A_{n-k}(b-1) b^k a^{n-k}.
$$

Taking $b = 1$ in Corollary 3 gives us the following identities.

Corollary 4. *For $n \geq 0$, $a \in \mathbb{N}$ with $a \equiv 1$ (mod 2), we have*

$$E_n(a) = \sum_{k=0}^{n} \binom{n}{k} E_k A_{n-k}(a-1)a^k$$

$$= \sum_{k=0}^{n} \binom{n}{k} E_k a^k \sum_{l=0}^{a-1} (-1)^l (2l+1)^{n-k}.$$

From (38), we have

$$J(a,b) = \frac{e^{abxt}}{2} \int_{\mathbb{Z}_p} e^{(2ax_1+a)t} d\mu_{-1}(x_1) \, \frac{2 \int_{\mathbb{Z}_p} e^{(2bx_2+b)t} d\mu_{-1}(x_2)}{\int_{\mathbb{Z}_p} e^{2abxt} d\mu_{-1}(x)}$$

$$= \frac{e^{abxt}}{2} \int_{\mathbb{Z}_p} e^{(2ax_1+a)t} d\mu_{-1}(x_1) \, 2 \sum_{l=0}^{a-1} (-1)^l e^{(2l+1)bt}$$

$$= \sum_{l=0}^{a-1} (-1)^l \int_{\mathbb{Z}_p} e^{(2x_1+1+bx+(2l+1)\frac{b}{a})at} d\mu_{-1}(x_1)$$

$$= \sum_{n=0}^{\infty} a^n \sum_{l=0}^{a-1} (-1)^l E_n\big(bx + (2l+1)\frac{b}{a}\big) \frac{t^n}{n!}, \tag{41}$$

where $a, b \in \mathbb{N}$ with $a \equiv 1$ (mod 2) and $b \equiv 1$ (mod 2). Interchanging a and b, we get

$$J(b,a) = \sum_{n=0}^{\infty} b^n \sum_{l=0}^{b-1} (-1)^l E_n\big(ax + (2l+1)\frac{a}{b}\big) \frac{t^n}{n!}. \tag{42}$$

As $J(a,b) = J(b,a)$, by (41) and (42), we obtain the following theorem.

Theorem 4. *For $n \geq 0$, $a, b \in \mathbb{N}$ with $a \equiv 1$ (mod 2) and $b \equiv 1$ (mod 2), we have*

$$a^n \sum_{l=0}^{a-1} (-1)^l E_n\big(bx + (2l+1)\frac{b}{a}\big) = b^n \sum_{l=0}^{b-1} (-1)^l E_n\big(ax + (2l+1)\frac{a}{b}\big).$$

Let us take $a = 1$ in Theorem 4. Then, we have

$$E_n(bx + b) = b^n \sum_{l=0}^{b-1} (-1)^l E_n\big(x + (2l+1)\frac{1}{b}\big).$$

Example 2. *Here, we illustrate Theorem 2 in the case of $n = 2$, $a = 7$, and $b = 3$. First, we note that $E_2(x) = x^2 - 1$. This follows from $E_2^*(x) = x^2 - x$ and the relation $E_n(x) = 2^n E_n^*(\frac{x+1}{2})$ that can be deduced from (11) and (14). Here, we need to show that*

$$\sum_{l=0}^{6} (-1)^l \left\{ (3x + \frac{3}{7}(2l+1))^2 - 1 \right\} = (\frac{3}{7})^2 \sum_{l=0}^{2} (-1)^l \left\{ (7x + \frac{7}{3}(2l+1))^2 - 1 \right\}. \tag{43}$$

Indeed, we can easily check that both the left- and right-hand side of (43) are equal to $9x^2 + 18x + \frac{824}{49}$.

3. Further Remarks

For $s \in \mathbb{C}$, the Riemann zeta function is defined by

$$\zeta(s) = \sum_{n=1}^{\infty} \frac{1}{n^s}, \quad (Re(s) > 1),$$

(see [14–16]).

It is well known that

$$\zeta(2n) = (-1)^{n-1} \frac{2^{2n-1}}{(2n)!} \pi^{2n} B_{2n}, \quad (n \geq 0), \tag{44}$$

(see [14,16]).

By (44), we get

$$
\begin{aligned}
z \cot(z) &= z \frac{\cos(z)}{\sin(z)} \\
&= z \frac{\frac{e^{iz} + e^{-iz}}{2}}{\frac{e^{iz} - e^{-iz}}{2i}}, \quad (i = \sqrt{-1}) \\
&= iz \left(1 + \frac{2}{e^{2iz} - 1} \right) \\
&= iz + \sum_{k=0}^{\infty} B_k \frac{(2iz)^k}{k!} \\
&= 1 + \sum_{k=1}^{\infty} \frac{B_{2k}}{(2k)!} 2^{2k} i^{2k} z^{2k} \\
&= 1 - 2 \sum_{k=1}^{\infty} \frac{\zeta(2k)}{\pi^{2k}} z^{2k} \\
&= 1 - 2 \sum_{n=1}^{\infty} \left(\sum_{k=1}^{\infty} \frac{z^{2k}}{(n\pi)^{2k}} \right) \\
&= 1 - 2 \sum_{n=1}^{\infty} \left(\frac{z}{n\pi} \right)^2 \left(1 - \left(\frac{z}{n\pi} \right)^2 \right)^{-1}.
\end{aligned}
\tag{45}
$$

Thus, by (45), we get

$$\cot(z) - \frac{1}{z} = -\sum_{n=1}^{\infty} \frac{2z}{(n\pi)^2} \left(1 - \left(\frac{z}{n\pi} \right)^2 \right)^{-1}. \tag{46}$$

From (39), we easily note that

$$\frac{d}{dz} \left(\log(\sin(z)) - \log(z) \right) = \sum_{n=1}^{\infty} \frac{d}{dz} \left(\log \left(1 - \left(\frac{z}{n\pi} \right)^2 \right) \right). \tag{47}$$

By (47), we easily get

$$\frac{\sin(z)}{z} = \prod_{n=1}^{\infty} \left(1 - \left(\frac{z}{n\pi} \right)^2 \right). \tag{48}$$

It is not difficult to show that

$$z \cot(z) - 2z \cot(2z) = z \tan(z). \tag{49}$$

From (45) and (49), we have

$$z \tan(z) = z \cot(z) - 2z \cot(2z)$$

$$= 2 \sum_{n=1}^{\infty} \left(\frac{2z}{n\pi} \right)^2 \left(1 - \left(\frac{2z}{n\pi} \right)^2 \right)^{-1} - 2 \sum_{n=1}^{\infty} \left(\frac{z}{n\pi} \right)^2 \left(1 - \left(\frac{z}{n\pi} \right)^2 \right)^{-1}. \tag{50}$$

By (50), we get

$$\frac{d}{dz} \left(-\log(\cos(z)) \right) = - \sum_{n=1}^{\infty} \frac{d}{dz} \left(\log \left(1 - \frac{4z^2}{(n\pi)^2} \right) \right) + \sum_{n=1}^{\infty} \frac{d}{dz} \left(\log \left(1 - \left(\frac{z}{n\pi} \right)^2 \right) \right). \tag{51}$$

Thus, from (51), we have

$$\sec(z) = \prod_{n=1}^{\infty} \left(\frac{1 - \left(\frac{z}{n\pi} \right)^2}{1 - \left(\frac{2z}{n\pi} \right)^2} \right) = \prod_{n=1}^{\infty} \left(1 - \left(\frac{2z}{(2n-1)\pi} \right)^2 \right)^{-1}, \tag{52}$$

which is equivalent to

$$\cos(z) = \prod_{n=1}^{\infty} \left(1 - \left(\frac{2z}{(2n-1)\pi} \right)^2 \right). \tag{53}$$

A random variable has the Laplace distribution with positive parameter μ and b if its probability density function is

$$f(x|\mu, b) = \frac{1}{2b} \exp \left(-\frac{|x - \mu|}{b} \right), \tag{54}$$

(see [17]).

The shorthand notation $X \sim \text{Laplace}(\mu, b)$ is used to indicate that the random variable X has the Laplace distribution with positive parameters μ and b. If $\mu = 0$ and $b = 1$, the positive half-time is exactly an exponential scaled by $\frac{1}{2}$.

We assume that the independent random variables $X_1, X_2, X_3, \cdots$ have the Laplace distribution with parameters 0 and 1, (i.e., $X_k \sim \text{Laplace}(0, 1)$, $k \in \mathbb{N}$). Let us put

$$Y = \sum_{k=1}^{\infty} \frac{X_k}{(2k - 1)\pi}. \tag{55}$$

Then, the characteristic function of Y is given by

$$\begin{aligned}
\sum_{n=0}^{\infty} E[Y^n] \frac{(2it)^n}{n!} &= E \left[\sum_{n=0}^{\infty} Y^n \frac{(2it)^n}{n!} \right] \\
&= E[e^{2iYt}] \\
&= E \left[e^{(\sum_{k=1}^{\infty} \frac{X_k}{(2k-1)\pi})2it} \right] \\
&= \prod_{k=1}^{\infty} E \left[e^{\frac{X_k}{(2k-1)\pi} 2it} \right].
\end{aligned} \tag{56}$$

Now, we observe that

$$
\begin{aligned}
E\left[e^{\frac{X_k}{(2k-1)\pi}2it}\right] &= \int_{-\infty}^{\infty} \frac{1}{2} e^{\left(\frac{2it}{(2k-1)\pi}\right)x} e^{-|x|}dx \\
&= \frac{1}{2}\int_{-\infty}^{0} e^{\left(\frac{2it}{(2k-1)\pi}\right)x} e^{x}dx + \frac{1}{2}\int_{0}^{\infty} e^{\left(\frac{2it}{(2k-1)\pi}\right)x} e^{-x}dx \\
&= \frac{1}{2}\frac{1}{1+\frac{2it}{(2k-1)\pi}} + \frac{1}{2}\frac{1}{1-\frac{2it}{(2k-1)\pi}} \\
&= \left(1+\left(\frac{2t}{(2k-1)\pi}\right)^2\right)^{-1}.
\end{aligned}
\tag{57}
$$

By (53), (56) and (57), we get

$$
\begin{aligned}
\sum_{n=0}^{\infty} E[Y^n]\frac{(2it)^n}{n!} &= \prod_{k=1}^{\infty} E\left[e^{\left(\frac{X_k}{(2k-1)\pi}\right)2it}\right] \\
&= \prod_{k=1}^{\infty}\left(1+\left(\frac{2t}{(2k-1)\pi}\right)^2\right)^{-1} \\
&= \frac{2}{e^t+e^{-t}} \\
&= \sum_{n=0}^{\infty} E_n \frac{t^n}{n!}.
\end{aligned}
\tag{58}
$$

Therefore, by comparing the coefficients on both sides of (58), we get

$$
2^n i^n E[Y^n] = E_n, \quad (n \geq 0).
\tag{59}
$$

Now, we assume that

$$
Z = \sum_{k=1}^{\infty} \frac{X_k}{2k\pi}.
\tag{60}
$$

Then, the characteristic function of Z is given by

$$
\begin{aligned}
\sum_{n=0}^{\infty} E[Z^n]\frac{(it)^n}{n!} &= E\left[\sum_{n=0}^{\infty} Z^n\frac{(it)^n}{n!}\right] \\
&= E[e^{Zit}] \\
&= E\left[e^{\sum_{k=1}^{\infty}\left(\frac{X_k}{2k\pi}\right)it}\right] \\
&= \prod_{k=1}^{\infty} E\left[e^{\left(\frac{X_k}{2k\pi}\right)it}\right].
\end{aligned}
\tag{61}
$$

Now, we note that

$$
\begin{aligned}
E\left[e^{\left(\frac{X_k}{2k\pi}\right)it}\right] &= \frac{1}{2}\int_{-\infty}^{\infty} e^{\left(\frac{it}{2k\pi}\right)x} e^{-|x|}\,dx \\
&= \frac{1}{2}\int_{-\infty}^{0} e^{\left(\frac{it}{2k\pi}\right)x} e^{x}\,dx + \frac{1}{2}\int_{0}^{\infty} e^{\left(\frac{it}{2k\pi}\right)x} e^{-x}\,dx \\
&= \frac{1}{2}\left(\frac{1}{1+\frac{it}{2k\pi}}\right) + \frac{1}{2}\left(\frac{1}{1-\frac{it}{2k\pi}}\right) \\
&= \frac{1}{1+\left(\frac{t}{k\pi}\right)^2}.
\end{aligned}
\tag{62}
$$

From (61) and (62), we have

$$
\begin{aligned}
\sum_{n=0}^{\infty} E[Z^n]\frac{(it)^n}{n!} &= \prod_{k=1}^{\infty} E\left[e^{\left(\frac{X_k}{2k\pi}\right)it}\right] \\
&= \prod_{k=1}^{\infty}\left(1+\left(\frac{t}{2k\pi}\right)^2\right)^{-1}.
\end{aligned}
\tag{63}
$$

On the other hand, by (48), we get

$$
\prod_{n=1}^{\infty}\left(1+\left(\frac{t}{n\pi}\right)^2\right)^{-1} = \frac{it}{\sin(it)} = \frac{2t}{e^t - e^{-t}}.
\tag{64}
$$

By replacing t by $\frac{t}{2}$, we have

$$
\begin{aligned}
\prod_{n=1}^{\infty}\left(1+\left(\frac{t}{2n\pi}\right)^2\right)^{-1} &= \frac{t}{e^{\frac{t}{2}} - e^{-\frac{t}{2}}} \\
&= \sum_{n=0}^{\infty}\left(\frac{1}{2}\right)^{n-1} b_n \frac{t^n}{n!}.
\end{aligned}
\tag{65}
$$

Therefore, by (63) and (65), we obtain the following equation

$$
i^n E[Z^n] = \left(\frac{1}{2}\right)^{n-1} b_n, \quad (n \geq 0).
\tag{66}
$$

4. Conclusions

In this paper, we obtained several identities of symmetry for the type 2 Bernoulli and Euler polynomials (see Theorems 1–4). Indeed, they are symmetric identities involving type 2 Bernoulli polynomials and power sums of consecutive odd positive integers, and the ones involving type 2 Euler polynomials and alternating power sums of odd positive integers. For the derivation of those identities, we introduced certain quotients of bosonic p-adic and fermionic p-adic integrals on $\mathbb{Z}_p$, which have built-in symmetries. We note that this idea of using certain quotients of p-adic integrals has produced abundant symmetric identities (see [5,7,8,18–21] and references therein).

We emphasize here that, even though there have been many results on symmetric identities relating to some special numbers and polynomials, this paper is the first one that deals with symmetric identities involving type 2 Bernoulli polynomials, type 2 Euler polynomials, power sums of odd positive integers and alternating power sums of odd positive integers.

In [22,23], we derived some identities involving special numbers and moments of random variables by using the generating functions of the moments of certain random variables. The related special numbers are Stirling numbers of the first and second kinds, degenerate Stirling numbers of the first and second kinds, derangement numbers, higher-order Bernoulli numbers and Bernoulli numbers of the second kind.

In this paper, we considered two random variables created from random variables having Laplace distributions and showed that their moments are closely connected with the type 2 Bernoulli and Euler numbers. Again, this is the first paper that interprets the type 2 Bernoulli and Euler numbers as the moments of certain random variables.

Author Contributions: Conceptualization, T.K.; Formal analysis, D.S.K. and T.K.; Funding acquisition, D.K.; Investigation, D.S.K., H.Y.K., D.K. and T.K.; Methodology, D.S.K. and T.K.; Project administration, D.K. and T.K.; Supervision, D.S.K. and T.K.; Validation, H.Y.K. and D.K.; Writing—original draft, T.K.; Writing—review and editing, D.S.K., H.Y.K. and D.K.

Funding: This work was supported by the National Research Foundation of Korea (NRF) grant funded by the Korea government (MSIT) (No. 2019R1C1C1003869).

Conflicts of Interest: The authors declare no conflict of interest.

References

1. Kim, Y.-H.; Hwang, K.-W. Symmetry of power sum and twisted Bernoulli polynomials. *Adv. Stud. Contemp. Math. (Kyungshang)* **2009**, *18*, 127–133.

2. Simsek, Y. Identities on the Changhee numbers and Apostol-type Daehee polynomials. *Adv. Stud. Contemp. Math. (Kyungshang)* **2017**, *27*, 199–212.

3. Jang, G.-W.; Kim, T. A note on type 2 degenerate Euler and Bernoulli polynomials. *Adv. Stud. Contemp. Math. (Kyungshang)* **2019**, *29*, 147–159.

4. Kim, T.; Kim, D.S. A note on type 2 Changhee and Daehee polynomials. *RACSAM* **2019**. [CrossRef]

5. Kim, D.S.; Lee, N.; Na, J.; Park, K.H. Abundant symmetry for higher-order Bernoulli polynomials (I). *Adv. Stud. Contemp. Math. (Kyungshang)* **2013**, *23*, 461–482.

6. Kim, T. q-Volkenborn integration. *Russ. J. Math. Phys.* **2002**, *9*, 288–299.

7. Kim, D.S. Symmetry identities for generalized twisted Euler polynomials twisted by unramified roots of unity. *Proc. Jangjeon Math. Soc.* **2012**, *15*, 303–316.

8. Kim, D.S.; Lee, N.; Na, J.; Park, K.H. Identities of symmetry for higher-order Euler polynomials in three variables (I). *Adv. Stud. Contemp. Math. (Kyungshang)* **2012**, *22*, 51–74. [CrossRef]

9. Kim, T. A Symmetry of power sum polynomials and multivariate fermionic p-adic invariant integral on $\mathbb{Z}_p$. *Russ. J. Math. Phys.* **2009**, *16*, 93–96. [CrossRef]

10. Kim, T. Some identities on the q-Euler polynomials of higher order and q-Stirling numbers by the fermionic p-adic integral on $\mathbb{Z}_p$. *Russ. J. Math. Phys.* **2009**, *16*, 484–491. [CrossRef]

11. Bagarello, F.; Trapani, C.; Triolo, S. Representable states on quasilocal quasi *-algebras. *J. Math. Phys.* **2011**, *52*, 013510. [CrossRef]

12. Bongiorno, B.; Trapani, C.; Triolo, S. Extensions of positive linear functionals on a topological *-algebra. *Rocky Mt. J. Math.* **2010**, *40*, 1745–1777. [CrossRef]

13. Trapani, C.; Triolo, S. Representations of modules over a *-algebra and related seminorms. *Stud. Math.* **2008**, *184*, 133–148. [CrossRef]

14. Kim, T. Euler numbers and polynomials associated with zeta functions. *Abstr. Appl. Anal.* **2008**, *52*, 581582. [CrossRef]

15. Andrews, G.E.; Askey, R.; Roy, R. Special functions. In *Encyclopedia of Mathematics and its Applications*; Cambridge University Press: Cambridge, UK, 1999.

16. Whittaker, E.T.; Watson, G.N. *A Course of Modern Analysis. an Introduction to the General Theory of Infinite Processes and of Analytic Functions; With an Account of the Principal Transcendental Functions*; Reprint of the 4th (1927) edition; Cambridge Mathematical Library: Cambridge, UK, 2017.

17. Kotz, S.; Kozubowski, T.J.; Podgórski, K. The Laplace distribution and generalizations. In *A Revisit with Applications to Communications, Economics, Engineering, and Finance*; Birkhäuser Boston, Inc.: Boston, MA, USA, 2001.

18. Kwon, J.; Sohn, G.; Park, J.-W. Symmetric identities for (h, q)-extensions of the generalized higher order modified q-Euler polynomials. *J. Comput. Anal. Appl.* **2018**, *24*, 1431–1438.

19. Kwon, J.; Park, J.-W. A note on symmetric identities for twisted Daehee polynomials. *J. Comput. Anal. Appl.* **2018**, *24*, 636–643.

20. Jeong, J.; Kang, D.-J.; Rim, S.-H. Symmetry identities of Changhee polynomials of type two. *Symmetry* **2018**, *10*, 740. [CrossRef]

21. Kim, T.; Kim, D.S.; Jang, G.-W.; Kwon, J. Symmetric Identities for Fubini Polynomials. *Symmetry* **2018**, *10*, 219. [CrossRef]

22. Kim, T.; Yao, Y.; Kim, D.S.; Kwon, H.-I. Some identities involving special numbers and moments of random variables. *To appear Rocky Mountain J. Math.* Available online: https://projecteuclid.org/euclid.rmjm/1538208042 (accessed on 30 April 2019).

23. Kim, T.; Kim, D.S. Extended Stirling numbers of the first kind associated with Daehee numbers and polynomials. *Rev. R. Acad. Cienc. Exactas Fís. Nat. Ser. A Mat. RACSAM* **2019**, *113*, 1159–1171. [CrossRef]

Article

Extended Degenerate r-Central Factorial Numbers of the Second Kind and Extended Degenerate r-Central Bell Polynomials

Dae San Kim [1], Dmitry V. Dolgy [2], Taekyun Kim [3] and Dojin Kim [4,*]

[1]   Department of Mathematics, Sogang University, Seoul 04107, Korea; dskim@sogang.ac.kr
[2]   Kwangwoon Institute for Advanced Studies, Kwangwoon University, Seoul 01897, Korea; d_dol@mail.ru
[3]   Department of Mathematics, Kwangwoon University, Seoul 01897, Korea; tkkim@kw.ac.kr
[4]   Department of Mathematics, Pusan National University, Busan 46241, Korea
*   Correspondence: kimdojin@pusan.ac.kr

Received: 2 March 2019; Accepted: 22 April 2019; Published: 24 April 2019

Abstract: In this paper, we introduce the extended degenerate r-central factorial numbers of the second kind and the extended degenerate r-central Bell polynomials. They are extended versions of the degenerate central factorial numbers of the second kind and the degenerate central Bell polynomials, and also degenerate versions of the extended r-central factorial numbers of the second kind and the extended r-central Bell polynomials, all of which have been studied by Kim and Kim. We study various properties and identities concerning those numbers and polynomials and also their connections.

Keywords: extended degenerate r-central factorial numbers of the second kind; extended degenerate r-central bell polynomials

1. Introduction

For $\lambda \in \mathbb{R}$, we recall that the degenerate exponential function $e_\lambda^x(t)$ is defined by (see [1–7])

$$e_\lambda^x(t) = (1 + \lambda t)^{\frac{x}{\lambda}} \tag{1}$$

When $x = 1$, we let $e_\lambda(t) = e_\lambda^1(t)$. Note that $\lim_{\lambda \to 0} e_\lambda^x(t) = e^{xt}$.

We use the notation $(x)_n$ to denote the falling factorial sequence $(x)_n$, which is defined by (see [8–14])

$$(x)_0 = 1, \quad (x)_n = x(x-1)\cdots(x-n+1), \quad (n \geq 1) \tag{2}$$

More generally, for $\lambda \in \mathbb{R}$, the λ-falling factorial sequence $(x)_{n,\lambda}$ is given by (see [4])

$$(x)_{0,\lambda} = 1, \quad (x)_{n,\lambda} = x(x-\lambda)(x-2\lambda)\cdots(x-(n-1)\lambda), \quad (n \geq 1) \tag{3}$$

Obviously, it is noted that $\lim_{\lambda \to 1}(x)_{n,\lambda} = (x)_n$, $\lim_{\lambda \to 0}(x)_{n,\lambda} = x^n$, $(n \geq 0)$.

In Reference [4], the λ-binomial expansion is defined by

$$(1 + \lambda t)^{\frac{x}{\lambda}} = \sum_{l=0}^{\infty} \binom{x}{l}_\lambda t^l = \sum_{l=0}^{\infty} (x)_{l,\lambda} \frac{t^l}{l!}, \tag{4}$$

where

$$\binom{x}{l}_{\lambda} = \frac{(x)_{l,\lambda}}{l!} = \frac{x(x-\lambda)(x-2\lambda)\cdots(x-(l-1)\lambda)}{l!}.$$

The central factorial sequence is given by

$$x^{[0]} = 1, \quad x^{[n]} = x\left(x+\frac{n}{2}-1\right)\left(x+\frac{n}{2}-2\right)\cdots\left(x-\frac{n}{2}+1\right), \quad (n \geq 1).$$

One can then easily show that the generating function of central factorial $x^{[n]}$, $(n \geq 0)$, is given by (see [3,15–20])

$$\left(\frac{t}{2}+\sqrt{1+\frac{t^2}{4}}\right)^{2x} = \sum_{n=0}^{\infty} x^{[n]}\frac{t^n}{n!} \tag{5}$$

As is defined in [18], for any non-negative integer n, the central factorial numbers of the first kind are given by

$$x^{[n]} = \sum_{k=0}^{n} t(n,k)x^k. \tag{6}$$

Then, from (5) and (6), we can show that the generating function of $t(n,k)$ satisfies the following equation:

$$\frac{1}{k!}\left(2\log\left(\frac{t}{2}+\sqrt{1+\frac{t^2}{4}}\right)\right)^k = \sum_{n=k}^{\infty} t(n,k)\frac{t^n}{n!}.$$

As the inverse to the central factorial numbers of the first kind, the central factorial numbers of the second kind are defined by (see [18,20–22])

$$x^n = \sum_{k=0}^{n} T_2(n,k)x^{[k]}, \quad (n \geq 0) \tag{7}$$

The generating function of $T_2(n,k)$ can be easily derived from (7), which is given by (see [18])

$$\frac{1}{k!}\left(e^{\frac{t}{2}}-e^{-\frac{t}{2}}\right)^k = \sum_{n=k}^{\infty} T_2(n,k)\frac{t^n}{n!}, \quad (k \geq 0) \tag{8}$$

It can immediately be seen from (8) that

$$k!T_2(n,k) = \sum_{j=0}^{k} \binom{k}{j}(-1)^j\left(\frac{1}{2}k-j\right)^n. \tag{9}$$

In Reference [22] were introduced the central Bell polynomials defined by

$$e^{x\left(e^{\frac{t}{2}}-e^{-\frac{t}{2}}\right)} = \sum_{n=0}^{\infty} B_n^{(c)}(x)\frac{t^n}{n!}. \tag{10}$$

The Dobinski-like formula for $B_n^{(c)}(x)$ is given by (see [22])

$$B_n^{(c)}(x) = \sum_{l=0}^{\infty}\sum_{k=0}^{\infty} \binom{l+k}{k}(-1)^k\frac{1}{(l+k)!}\left(\frac{l}{2}-\frac{k}{2}\right)^{l+1} \tag{11}$$

In Reference [3], the degenerate central factorial polynomials of the second kind are defined by

$$\frac{1}{k!}\left(e_\lambda^{\frac{1}{2}}(t) - e_\lambda^{-\frac{1}{2}}(t)\right)^k e_\lambda^x(t) = \sum_{n=k}^\infty T_{2,\lambda}(n,k|x)\frac{t^n}{n!}, \quad (k \geq 0). \tag{12}$$

When $x = 0$, $T_{2,\lambda}(n,k) = T_{2,\lambda}(n,k|0)$, these are called degenerate central factorial numbers of the second kind.

Let us recall that the degenerate central Bell polynomials are defined by (see [3])

$$e^{x\left(e_\lambda^{\frac{1}{2}}(t) - e_\lambda^{-\frac{1}{2}}(t)\right)} = \sum_{n=0}^\infty B_{n,\lambda}^{(c)}(x)\frac{t^n}{n!}, \tag{13}$$

In particular, $B_{n,\lambda}^{(c)} = B_{n,\lambda}^{(c)}(1)$ are called the degenerate central Bell numbers.

Note that $\lim_{\lambda \to 0} B_{n,\lambda}^{(c)}(x) = B_n^{(c)}(x)$, $(n \geq 0)$.

Carlitz [1] introduced the degenerate Stirling, Bernoulli, and Eulerian numbers as the first degenerate special numbers. Broder [23] investigated the r-Stirling numbers of the first and second kind as the numbers counting restricted permutations and restricted partitions, respectively. We recall here that the r-Stirling numbers of the second kind are given by (see [23])

$$\frac{1}{k!}e^{rt}(e^t - 1)^k = \sum_{n=k}^\infty S_2^{(r)}(n+r,k+r)\frac{t^n}{n!}, \tag{14}$$

In this paper, we will introduce the extended degenerate r-central factorial numbers of the second kind and the extended degenerate r-central Bell polynomials. Central analogues of Stirling numbers of the second kind and Bell polynomials are, respectively, the central factorial numbers of the second kind and the central Bell polynomials. Degenerate versions of the central factorial numbers of the second kind and the central Bell polynomials are, respectively, the degenerate central factorial numbers of the second kind and the degenerate central Bell polynomials. Extended versions of the degenerate central factorial numbers of the second kind and the degenerate central Bell polynomials are, respectively, the extended degenerate r-central factorial numbers of the second kind and the extended degenerate r-central Bell polynomials. The central factorial numbers of the second kind have many applications in such diverse areas as approximation theory [21], finite difference calculus, spline theory, spectral theory of differential operators [24,25], and algebraic geometry [26,27]. For broad applications of the related complete and incomplete Bell polynomials, we let the reader consult the introduction in [11]. Here, we will study various properties and identities relating to those numbers and polynomials, and also their connections. Finally, we note that the present paper can be useful in the area of non-integer systems and let the reader refer to [28] for more research in this direction.

2. Extended Degenerate r-Central Factorial Numbers of the Second Kind and Extended Degenerate r-Central Bell Polynomials

From (12) and (13), we note that

$$\sum_{n=0}^\infty B_{n,\lambda}^{(c)}(x)\frac{t^n}{n!} = \sum_{k=0}^\infty x^k \sum_{n=k}^\infty T_{2,\lambda}(n,k)\frac{t^n}{n!}$$
$$= \sum_{n=0}^\infty \sum_{k=0}^n x^k T_{2,\lambda}(n,k)\frac{t^n}{n!}. \tag{15}$$

One can compare the coefficients on both sides of (15) to obtain

$$B_{n,\lambda}^{(c)}(x) = \sum_{k=0}^{n} T_{2,\lambda}(n,k)x^k, \quad (n \geq 0). \tag{16}$$

Throughout this paper, we assume that r is a nonnegative integer. The following definition is motivated by (14).

Definition 1. *The extended degenerate r-central factorial numbers of the second kind $T_{\lambda}^{(r)}(n+r,k+r)$ are defined as*

$$\frac{1}{k!}e_{\lambda}^{r}(t)\left(e_{\lambda}^{\frac{1}{2}}(t) - e_{\lambda}^{-\frac{1}{2}}(t)\right)^k = \sum_{n=k}^{\infty} T_{\lambda}^{(r)}(n+r,k+r)\frac{t^n}{n!}. \tag{17}$$

Note that $\lim\limits_{\lambda \to 0} T_{\lambda}^{(r)}(n+r,k+r) = T^{(r)}(n+r,k+r), \quad (n,k \geq 0),$

where $T^{(r)}(n+r,k+r)$ is the extended r-central factorial numbers of the second kind given by

$$\frac{1}{k!}e^{rt}\left(e^{\frac{t}{2}} - e^{-\frac{t}{2}}\right)^k = \sum_{n=k}^{\infty} T^{(r)}(n+r,k+r)\frac{t^n}{n!}. \tag{18}$$

Theorem 1. *For n, $k \in \mathbb{N} \cup \{0\}$, with $n \geq k$, we have*

$$T_{\lambda}^{(r)}(n+r,k+r) = \sum_{l=k}^{n} \binom{n}{l} T_{2,\lambda}(l,k)(r)_{n-l,\lambda}.$$

Proof. By (17), we get

$$\begin{aligned}
\frac{1}{k!}\left(e_{\lambda}^{\frac{1}{2}}(t) - e_{\lambda}^{-\frac{1}{2}}(t)\right)^k e_{\lambda}^{r}(t) &= \sum_{l=k}^{\infty} T_{2,\lambda}(l,k)\frac{t^l}{l!} \sum_{m=0}^{\infty} (r)_{m,\lambda}\frac{t^m}{m!} \\
&= \sum_{n=k}^{\infty}\sum_{l=k}^{n} \binom{n}{l} T_{2,\lambda}(l,k)(r)_{n-l,\lambda}\frac{t^n}{n!}.
\end{aligned} \tag{19}$$

Therefore, by (17) and (19), we obtain the result. $\square$

We note that by taking the limit as λ tends to 0, we get

$$T^{(r)}(n+r,k+r) = \sum_{l=k}^{n} \binom{n}{l} r^{n-l} T_2(l,k). \tag{20}$$

Theorem 2. *For n, $k \geq 0$, with $n \geq k$, we have*

$$T_{\lambda}^{(r)}(n+r,k+r) = \sum_{m=k}^{n}\sum_{l=k}^{m} \binom{m}{l} S_1(n,m)T_2(l,k)\lambda^{n-m}r^{m-l}, \tag{21}$$

where $S_1(n,m)$ are the signed Stirling numbers of the first kind.

Proof. Replacing t by $\frac{1}{\lambda}\log(1+\lambda t)$ in (18), we obtain

$$\frac{1}{k!}\left(e_\lambda^{\frac{1}{2}}(t)-e_\lambda^{-\frac{1}{2}}(t)\right)^k e_\lambda^r(t) = \sum_{m=k}^\infty \lambda^{-m}T^{(r)}(m+r,k+r)\frac{1}{m!}\left(\log(1+\lambda t)\right)^m$$

$$= \sum_{m=k}^\infty \lambda^{-m}T^{(r)}(m+r,k+r)\sum_{n=m}^\infty S_1(n,m)\frac{\lambda^n t^n}{n!} \tag{22}$$

$$= \sum_{n=k}^\infty\sum_{m=k}^n \lambda^{n-m}S_1(n,m)T^{(r)}(m+r,k+r)\frac{t^n}{n!}.$$

Now, by substituting the expression of $T^{(r)}(m+r,k+r)$ in (20) into (22), we finally get

$$\frac{1}{k!}\left(e_\lambda^{\frac{1}{2}}(t)-e_\lambda^{-\frac{1}{2}}(t)\right)^k e_\lambda^r(t) = \sum_{n=k}^\infty\sum_{m=k}^n\sum_{l=k}^m \binom{m}{l}S_1(n,m)T_2(l,k)\lambda^{n-m}r^{m-l}\frac{t^n}{n!},$$

from which the result follows. $\square$

Example 1. *Here, we will illustrate the formula* (21) *for small values of n. The following values of $T_2(n,k)$ can be determined, for example, from the formula in* (9):

$$T_2(n,n)=1,\ T_2(n,0)=\delta_{n,0},\ T_2(2,1)=T_2(3,2)=T_2(4,1)=T_2(4,3)=0, T_2(3,1)=\tfrac{1}{4},\ T(4,2)=1. \tag{23}$$

In addition, we recall the following values of $S_1(n,k)$:

$$S_1(n,n)=1,\ S_1(n,0)=\delta_{n,0},\ S_1(2,1)=-1,\ S_1(3,1)=2,$$
$$S_1(3,2)=-3,\ S_1(4,1)=S_1(4,3)=-6,\ S_1(4,2)=11. \tag{24}$$

Now, from (21)*,* (23)*, and* (24)*, we easily have*

$$T_\lambda^{(r)}(n+r,n+r)=1,\ T_\lambda^{(r)}(1+r,r)=r,\ T_\lambda^{(r)}(2+r,r)=-\lambda r+r^2,$$

$$T_\lambda^{(r)}(3+r,r)=2\lambda^2 r-3\lambda r^2+r^3,\ T_\lambda^{(r)}(4+r,r)=-6\lambda^3 r+11\lambda^2 r^2-6\lambda r^3+r^4,$$

$$T_\lambda^{(r)}(2+r,1+r)=-\lambda+2r,\ T_\lambda^{(r)}(3+r,1+r)=2\lambda^2-6\lambda r+3r^2+\frac{1}{4},$$

$$T_\lambda^{(r)}(3+r,2+r)=-3\lambda+3r,\ T_\lambda^{(r)}(4+r,1+r)=-6\lambda^3+22\lambda^2 r-18\lambda r^2-\frac{3}{2}\lambda+4r^3+r,$$

$$T_\lambda^{(r)}(4+r,2+r)=11\lambda^2-18\lambda r+6r^2+1,\ T_\lambda^{(r)}(4+r,3+r)=-6\lambda+4r.$$

Theorem 3. *For $n,\ k\geq 0$, with $n\geq k$, we have*

$$T_\lambda^{(r)}(n+r,k+r) = \sum_{m=0}^{n-k}\binom{m+k}{m}m!\binom{r}{m}T_{2,\lambda}\left(n,m+k\Big|\frac{m}{2}\right).$$

Proof. Now, we observe that

$$\frac{1}{k!}e_\lambda^r(t)\left(e_\lambda^{\frac{1}{2}}(t) - e_\lambda^{-\frac{1}{2}}(t)\right)^k = \frac{1}{k!}e_\lambda^{\frac{r}{2}}(t)\left(e_\lambda^{\frac{1}{2}}(t) - e_\lambda^{-\frac{1}{2}}(t) + e_\lambda^{-\frac{1}{2}}(t)\right)^r\left(e_\lambda^{\frac{1}{2}}(t) - e_\lambda^{-\frac{1}{2}}(t)\right)^k$$

$$= \frac{1}{k!}\sum_{m=0}^{\infty}\binom{r}{m}\left(e_\lambda^{\frac{1}{2}}(t) - e_\lambda^{-\frac{1}{2}}(t)\right)^{m+k}e_\lambda^{\frac{m}{2}}(t)$$

$$= \sum_{m=0}^{\infty}\binom{r}{m}\frac{(m+k)!}{k!}\frac{1}{(m+k)!}\left(e_\lambda^{\frac{1}{2}}(t) - e_\lambda^{-\frac{1}{2}}(t)\right)^{m+k}e_\lambda^{\frac{m}{2}}(t) \qquad (25)$$

$$= \sum_{m=0}^{\infty}\binom{r}{m}m!\binom{m+k}{m}\sum_{n=m+k}^{\infty}T_{2,\lambda}(n,m+k|\frac{m}{2})\frac{t^n}{n!}$$

$$= \sum_{n=k}^{\infty}\sum_{m=0}^{n-k}\binom{r}{m}m!\binom{m+k}{m}T_{2,\lambda}(n,m+k|\frac{m}{2})\frac{t^n}{n!}.$$

Therefore, by (17) and (25), we obtain the theorem. $\square$

One can easily show that the inverse function of $e_\lambda(t)$ is given by

$$\log_\lambda(t) = \frac{t^\lambda - 1}{\lambda}, \quad (t > 0),$$

so that $e_\lambda(\log_\lambda(t)) = \log_\lambda(e_\lambda(t)) = t$, $\lim_{\lambda \to 0}\log_\lambda(t) = \log(t)$.

If $g(t) = e_\lambda^{\frac{1}{2}}(t) - e_\lambda^{-\frac{1}{2}}(t)$, then one can see that

$$g^{-1}(t) = \log_\lambda\left(\frac{t}{2} + \sqrt{1 + \frac{t^2}{4}}\right)^2, \qquad (26)$$

where $g \circ g^{-1}(t) = g^{-1} \circ g(t) = t$.

Theorem 4. *For $n \geq 0$, we have*

$$(x+r)_{n,\lambda} = \sum_{k=0}^{n}T_\lambda^{(r)}(n+r, k+r)x^{[k]}$$

$$= \sum_{k=0}^{n}T_{2,\lambda}(n,k|\frac{k}{2} + r)(x)_k.$$

Proof. By (1) and (4), we get

$$
\begin{aligned}
e_\lambda^{x+r}(t) &= e_\lambda^r(t)\,(e_\lambda(t) - 1 + 1)^x \\
&= e_\lambda^r(t) \sum_{k=0}^{\infty} (x)_k \frac{1}{k!}\, (e_\lambda(t) - 1)^k \\
&= \sum_{k=0}^{\infty} (x)_k \frac{1}{k!} e_\lambda^{\frac{k}{2}+r}(t) \left(e_\lambda^{\frac{1}{2}}(t) - e_\lambda^{-\frac{1}{2}}(t) \right)^k \\
&= \sum_{k=0}^{\infty} (x)_k \sum_{n=k}^{\infty} T_{2,\lambda}\!\left(n,k\Big|\frac{k}{2}+r\right) \frac{t^n}{n!} \\
&= \sum_{n=0}^{\infty} \sum_{k=0}^{n} (x)_k\, T_{2,\lambda}\!\left(n,k\Big|\frac{k}{2}+r\right) \frac{t^n}{n!}.
\end{aligned}
\tag{27}
$$

Now, from the observations in (26) and (5), we have

$$
\begin{aligned}
e_\lambda^{x+r}(t) &= e_\lambda^r(t)e_\lambda^x(t) \\
&= e_\lambda^r(t) \left(e_\lambda \left(\log_\lambda \left(\frac{g(t)}{2} + \sqrt{1 + \frac{g(t)^2}{4}} \right)^2 \right) \right)^x \\
&= e_\lambda^r(t) \left(\frac{g(t)}{2} + \sqrt{1 + \frac{g(t)^2}{4}} \right)^{2x} \\
&= \sum_{k=0}^{\infty} x^{[k]} \frac{1}{k!} e_\lambda^r(t) \left(e_\lambda^{\frac{1}{2}}(t) - e_\lambda^{-\frac{1}{2}}(t) \right)^k \\
&= \sum_{k=0}^{\infty} x^{[k]} \sum_{n=k}^{\infty} T_\lambda^{(r)}(n+r, k+r) \frac{t^n}{n!} \\
&= \sum_{n=0}^{\infty} \sum_{k=0}^{n} x^{[k]} T_\lambda^{(r)}(n+r, k+r) \frac{t^n}{n!}.
\end{aligned}
\tag{28}
$$

From (4), we note also that

$$
e_\lambda^{x+r}(t) = \sum_{n=0}^{\infty} (x+r)_{n,\lambda} \frac{t^n}{n!}.
\tag{29}
$$

Therefore, by (27), (28), and (29), we have the desired result. $\square$

Note that, taking the limit as λ tends to 0, we have

$$
(x+r)^n = \sum_{k=0}^{n} T^{(r)}(n+r, k+r) x^{[k]} = \sum_{k=0}^{n} T_2\!\left(n,k\Big|\frac{k}{2}+r\right)(x)_k .
$$

Definition 2. *The extended degenerate r-central Bell polynomials* $B_{n,\lambda}^{(c,r)}(x)$ *are defined by*

$$
e_\lambda^r(t) e^{x\left(e_\lambda^{\frac{1}{2}}(t) - e_\lambda^{-\frac{1}{2}}(t) \right)} = \sum_{n=0}^{\infty} B_{n,\lambda}^{(c,r)}(x) \frac{t^n}{n!}.
\tag{30}
$$

Specifically, $B_{n,\lambda}^{(c,r)}(1) = B_{n,\lambda}^{(c,r)}$ *are called the extended degenerate r-central Bell numbers.*

Theorem 5. *For $n \geq 0$, we have*

$$B_{n,\lambda}^{(c,r)}(x) = \sum_{k=0}^{n} x^k T_\lambda^{(r)}(n+r,k+r).$$

Proof. From (30), we note that

$$
\begin{aligned}
e_\lambda^r(t) e^{x\left(e_\lambda^{\frac{1}{2}}(t) - e^{-\frac{1}{2}}(t)\right)}
&= \sum_{k=0}^{\infty} x^k \frac{1}{k!} \left(e_\lambda^{\frac{1}{2}}(t) - e^{-\frac{1}{2}}(t)\right)^k e_\lambda^r(t) \\
&= \sum_{k=0}^{\infty} x^k \sum_{n=k}^{\infty} T_\lambda^{(r)}(n+r,k+r)\frac{t^n}{n!} \\
&= \sum_{n=0}^{\infty} \sum_{k=0}^{n} x^k T_\lambda^{(r)}(n+r,k+r)\frac{t^n}{n!}.
\end{aligned}
\tag{31}
$$

Therefore, from (30) and (31), the theorem follows. $\square$

The central difference operator δ for a given function f is given by

$$\delta f(x) = f\left(x + \frac{1}{2}\right) - f\left(x - \frac{1}{2}\right),$$

and by induction we can show

$$\delta^k f(x) = \sum_{l=0}^{k} \binom{k}{l}(-1)^{k-l} f\left(x + l - \frac{k}{2}\right), \quad (k \geq 0). \tag{32}$$

Theorem 6. *Let n, k be nonnegative integers. Then, we have*

$$\frac{1}{k!}\delta^k(r)_{n,\lambda} = \begin{cases} 0, & \text{if } n < k, \\ T_\lambda^{(r)}(n+r,k+r), & \text{if } n \geq k. \end{cases}$$

Proof. By the binomial theorem, we have

$$
\begin{aligned}
\frac{1}{k!}e_\lambda^r(t)\left(e_\lambda^{\frac{1}{2}}(t) - e^{-\frac{1}{2}}(t)\right)^k
&= \frac{1}{k!}e_\lambda^{r-\frac{k}{2}}(t) \sum_{l=0}^{k} \binom{k}{l}(-1)^{k-l} e_\lambda^l(t) \\
&= \frac{1}{k!} \sum_{l=0}^{k} \binom{k}{l}(-1)^{k-l} e_\lambda^{r-\frac{k}{2}+l}(t) \\
&= \sum_{n=0}^{\infty} \frac{1}{k!} \sum_{l=0}^{k} \binom{k}{l}(-1)^{k-l}\left(r - \frac{k}{2} + l\right)_{n,\lambda}\frac{t^n}{n!}.
\end{aligned}
\tag{33}
$$

If we choose $f(x) = (x)_{n,\lambda}$, $(n \geq 0)$ in (32), then we have

$$\delta^k(r)_{n,\lambda} = \sum_{l=0}^{k} \binom{k}{l}\left(r + l - \frac{k}{2}\right)_{n,\lambda}(-1)^{k-l}. \tag{34}$$

From (33) and (34), the following equation is obtained.

$$\frac{1}{k!}e_\lambda^r(t)\left(e_\lambda^{\frac{1}{2}}(t) - e_\lambda^{-\frac{1}{2}}(t)\right)^k = \sum_{n=0}^{\infty}\frac{1}{k!}\delta^k(r)_{n,\lambda}\frac{t^n}{n!}. \tag{35}$$

Therefore, by (17) and (35), we have the result. $\square$

From Theorem 4 and Theorem 5, we have

$$\begin{aligned}
B_{n,\lambda}^{(c,r)}(x) &= \sum_{k=0}^{n} T_\lambda^{(r)}(n+r, k+r)x^k \\
&= \sum_{k=0}^{n} x^k \frac{1}{k!}\delta^k(r)_{n,\lambda}, \quad (n \geq 0).
\end{aligned} \tag{36}$$

Theorem 7. *For $n \geq 0$, we have*

$$B_{n,\lambda}^{(c,r)}(x) = \sum_{m=0}^{n}\binom{n}{m}(r)_{n-m,\lambda}B_{m,\lambda}^{(c)}(x).$$

Proof. From (30), we note that

$$\begin{aligned}
\sum_{n=0}^{\infty} B_{n,\lambda}^{(c,r)}(x)\frac{t^n}{n!} &= e_\lambda^r(t)e^{x\left(e_\lambda^{\frac{1}{2}}(t) - e_\lambda^{-\frac{1}{2}}(t)\right)} \\
&= \sum_{l=0}^{\infty}(r)_{l,\lambda}\frac{t^l}{l!}\sum_{m=0}^{\infty}B_{m,\lambda}^{(c)}\frac{t^m}{m!} \\
&= \sum_{n=0}^{\infty}\sum_{m=0}^{n}\binom{n}{m}(r)_{n-m,\lambda}B_{m,\lambda}^{(c)}(x)\frac{t^n}{n!}.
\end{aligned} \tag{37}$$

Therefore, by comparing the coefficients on both sides of (37), the desired result is achieved. $\square$

Theorem 8. *For m, n, $k \geq 0$, with $n \geq m+k$, we have*

$$\binom{m+k}{m}T_\lambda^{(r)}(n+r, m+k+r) = \sum_{l=m}^{n-k}\binom{n}{l}T_\lambda^{(r)}(l+r, m+r)T_{2,\lambda}(n-l, k).$$

Proof. We further observe that

$$\begin{aligned}
\frac{1}{m!}e_\lambda^r(t)\left(e_\lambda^{\frac{1}{2}}(t) - e_\lambda^{-\frac{1}{2}}(t)\right)^m \frac{1}{k!}\left(e_\lambda^{\frac{1}{2}}(t) - e_\lambda^{-\frac{1}{2}}(t)\right)^k &= \frac{(m+k)!}{m!k!}\frac{1}{(m+k)!}e_\lambda^r(t)\left(e_\lambda^{\frac{1}{2}}(t) - e_\lambda^{-\frac{1}{2}}(t)\right)^{m+k} \\
&= \binom{m+k}{m}\sum_{n=m+k}^{\infty}T_\lambda^{(r)}(n+r, m+k+r)\frac{t^n}{n!},
\end{aligned} \tag{38}$$

where m, k are nonnegative integers. Alternatively, the left-hand side of (38) can be expressed by

$$\begin{aligned}
\frac{1}{m!}e_\lambda^r(t)\left(e_\lambda^{\frac{1}{2}}(t) - e_\lambda^{-\frac{1}{2}}(t)\right)^m \frac{1}{k!}\left(e_\lambda^{\frac{1}{2}}(t) - e_\lambda^{-\frac{1}{2}}(t)\right)^k &= \sum_{l=m}^{\infty}T_\lambda^{(r)}(l+r, m+r)\frac{t^l}{l!}\sum_{j=k}^{\infty}T_{2,\lambda}(j, k)\frac{t^j}{j!} \\
&= \sum_{n=m+k}^{\infty}\sum_{l=m}^{n-k}\binom{n}{l}T_\lambda^{(r)}(l+r, m+r)T_{2,\lambda}(n-l, k)\frac{t^n}{n!}.
\end{aligned} \tag{39}$$

Therefore, by (38) and (39), the desired identity is obtained. $\square$

3. Conclusions

In recent years, many researchers have studied a lot of old and new special numbers and polynomials by means of generating functions, through combinatorial methods, umbral calculus, differential equations, p-adic integrals, p-adic q-integrals, special functions, complex analyses, and so on.

The study of degenerate versions of special numbers and polynomials began with Carlitz [1]. Kim and his colleagues have been studying degenerate versions of various special numbers and polynomials by making use of the same methods. Studying degenerate versions of known special numbers and polynomials can be very a fruitful research and is highly rewarding. For example, this line of study led even to the introduction of degenerate Laplace transforms and degenerate gamma functions (see [4]).

In this paper, we introduced the extended degenerate r-central factorial numbers of the second kind and the extended degenerate r-central Bell polynomials. We studied various properties and identities relating to those numbers and polynomials and also their connections. This study was done by using generating function techniques.

Central analogues of Stirling numbers of the second kind and Bell polynomials are, respectively, the central factorial numbers of the second kind and the central Bell polynomials. Degenerate versions of the central factorial numbers of the second kind and the central Bell polynomials are, respectively, the degenerate central factorial numbers of the second kind and the degenerate central Bell polynomials. Extended versions of the degenerate central factorial numbers of the second kind and the degenerate central Bell polynomials are, respectively, the extended degenerate r-central factorial numbers of the second kind and the extended degenerate r-central Bell polynomials. The central factorial numbers of the second kind have many applications in diverse areas such as approximation theory [21], finite difference calculus, spline theory, spectral theory of differential operators [24,25], and algebraic geometry [26,27].

For future research projects, we would like to continue to work on some special numbers and polynomials and their degenerate versions, as well as try to explore their applications not only in mathematics but also in the sciences and engineering [29].

Author Contributions: Conceptualization, D.S.K., T.K. and D.K.; Formal analysis, D.S.K., D.V.D., T.K. and D.K.; Investigation, D.S.K., D.V.D., T.K. and D.K.; Methodology, D.S.K., T.K. and D.K.; Supervision, D.S.K.; Writing—original draft, T.K.; Writing—review & editing, D.S.K., D.V.D., T.K. and D.K.

Funding: This work was supported by the National Research Foundation of Korea(NRF) grant funded by the Korea government(MSIT) (No. 2019R1C1C1003869).

References

1. Carlitz, L. Degenerate Stirling, Bernoulli and Eulerian numbers. *Util. Math.* **1979**, *15*, 51–88.
2. Jeong, J.; Rim, S.-H.; Kim, B.M. On finite-times degenerate Cauchy numbers and polynomials. *Adv. Differ. Equ.* **2015**, *2015*, 321. [CrossRef]
3. Kim, T.; Kim, D.S. Degenerate central Bell numbers and polynomials. *Rev. R. Acad. Clenc. Exactas Fis. Nat. Ser. A Mat. RACSAM* **2019**. [CrossRef]
4. Kim, T.; Kim, D.S. Degenerate Laplace transform and degenerate gamma function. *Russ. J. Math. Phys.* **2017**, *24*, 241–248. [CrossRef]
5. Kim, Y.; Kim, B.M.; Jang, L.-C.; Kwon, J. A note on modified degenerate gamma and Laplace transformation. *Symmetry* **2018**, *10*, 471. [CrossRef]
6. Pyo, S.-S. Degenerate Cauchy numbers and polynomials of the fourth kind. *Adv. Stud. Contemp. Math. (Kyungshang)* **2018**, *28*, 127–138.

7. Upadhyaya, L.M. On the degenerate Laplace transform IV. *Int. J. Eng. Sci. Res.* **2018**, *6*, 198–209.

8. Carlitz, L. Some remarks on the Bell numbers. *Fibonacci Quart.* **1980**, *18*, 66–73.

9. Duran, U.; Acikgoz, M.; Araci, S. On (q, r, w)-Stirling numbers of the second kind. *J. Inequal. Spec. Funct.* **2018**, *9*, 9–16.

10. Kim, T.; Yao, Y.; Kim, D.S.; Jang, G.-W. Degenerate r-Stirling numbers and r-Bell polynomials. *Russ. J. Math. Phys.* **2018**, *25*, 44–58. [CrossRef]

11. Kim, T.; Kim, D.S.; Kim, G.-W. On central complete and incomplete Bell polynomials I. *Symmetry* **2019**, *11*, 288. [CrossRef]

12. Roman, S. *The Umbral Calculus*; Pure and Applied Mathematics 111; Academic Press Inc. [Harcourt Brace Jovanovich, Publishers]: New York, NY, USA, 1984.

13. Simsek, Y. Identities and relations related to combinatorial numbers and polynomials. *Proc. Jangjeon Math. Soc.* **2017**, *20*, 127–135.

14. Simsek, Y. Identities on the Changhee numbers and Apostol-type Daehee polynomials. *Adv. Stud. Contemp. Math. (Kyungshang)* **2017**, *27*, 199–212.

15. Carlitz, L.; Riordan, J. The divided central differences of zero. *Can. J. Math.* **1963**, *15*, 94–100. [CrossRef]

16. Charalambides, C.A. Central factorial numbers and related expansions. *Fibonacci Quart.* **1981**, *19*, 451–456.

17. Kim, D.S.; Dolgy, D.V.; Kim, D.; Kim, T. Some identities on r-central factorial numbers and r-central Bell polynomials. *arXiv* **2019**, arXiv:1903.11689.

18. Kim, T. A note on central factorial numbers. *Proc. Jangjeon Math. Soc.* **2018**, *21*, 575–588.

19. Kim, T.; Kim, D.S.; Jang, G.-W.; Kwon, J. Extended central factorial polynomials of the second kind. *Adv. Differ. Equ.* **2019**, *2019*, 24. [CrossRef]

20. Zhang, W. Some identities involving the Euler and the central factorinal numbers. *Fibonacci Quart.* **1998**, *36*, 154–157.

21. Butzer, P.L.; Schmidt, M.; Stark, E.L.; Vogt, L. Central factorial numbers; their main properties and some applications. *Numer. Funct. Anal. Optim.* **1989**, *10*, 419–488. [CrossRef]

22. Kim, T.; Kim, D.S. A note on central Bell numbers and polynomials. *Russ. J. Math. Phys.* **2019**, to appear.

23. Broder, A.Z. The r-Stirling numbers. *Discret. Math.* **1984**, *49*, 241–259. [CrossRef]

24. Everitt, W.N.; Kwon, K.H.; Littlejohn, L.L.; Wellman, R.; Yoon, G.J. JacobiStirling numbers, Jacobi polynomials, and the left-definite analysis of the classical Jacobi differential expression. *J. Comput. Appl. Math.* **2007**, *208*, 29–56. [CrossRef]

25. Loureiro, A.F. New results on the Bochner condition about classical orthogonal polynomials. *J. Math. Anal. Appl.* **2010**, *364*, 307–323. [CrossRef]

26. Eastwood, M.; Goldschmidt, H. Zero-energy fields on complex projective space. *J. Differ. Geom.* **2013**, *94*, 129–157. [CrossRef]

27. Shadrin, S.; Spitz, L.; Zvonkine, D. On double Hurwitz numbers with completed cycles. *J. Lond. Math. Soc.* **2012**, *86*, 407–432. [CrossRef]

28. Caponetto, R.; Dongola, G.; Fortuna, L.; Gallo, A. New results on the synthesis of FO-PID controllers. *Commun. Nonlinear Sci. Numer. Simul.* **2010**, *15*, 997–1007. [CrossRef]

29. Kim, D.S.; Kim, T. A Note on Polyexponential and Unipoly Functions. *Russ. J. Math. Phys.* 2019, *94*, 40–49. [CrossRef]

 *symmetry*

Article

The Extended Minimax Disparity RIM Quantifier Problem

Dug Hun Hong

Department of Mathematics, Myongji University, Yongin Kyunggido 449-728, Korea; dhhong@mju.ac.kr

Received: 8 March 2019; Accepted: 1 April 2019; Published: 3 April 2019

Abstract: An interesting regular increasing monotone (RIM) quantifier problem is investigated. Amin and Emrouznejad [Computers & Industrial Engineering 50(2006) 312–316] have introduced the extended minimax disparity OWA operator problem to determine the OWA operator weights. In this paper, we propose a corresponding continuous extension of an extended minimax disparity OWA model, which is the extended minimax disparity RIM quantifier problem, under the given orness level and prove it analytically.

Keywords: fuzzy sets; RIM quantifier; extended minimax disparity; OWA model; RIM quantifier problem

---

1. Introduction

One of the important topic in the theory of ordered weighted averaging (OWA) operators is the determination of the associated weights. Several authors have suggested a number of methods for obtaining associated weights in various areas such as decision making, approximate reasoning, expert systems, data mining, fuzzy systems and control [1–18]. Researchers can easily see most of OWA papers in the recent bibliography published in Emrouznejad and Marra [5]. Yager [16] proposed RIM quantifiers as a method for finding OWA weight vectors through fuzzy linguistic quantifiers. Liu [19] and Liu and Da [20] gave solutions to the maximum-entropy RIM quantifier model when the generating functions are differentiable. Liu and Lou [21] studied the equivalence of solutions to the minimax ratio and maximum-entropy RIM quantifier models, and the equivalence of solutions to the minimax disparity and minimum-variance RIM quantifier problems. Hong [22,23] gave the proof of the minimax ratio RIM quantifier problem and the minimax disparity RIM quantifier model when the generating functions are absolutely continuous. He also gave solutions to the maximum-entropy RIM quantifier model and the minimum-variance RIM quantifier model when the generating functions are Lebesgue integrable. Liu [24] proposed a general RIM quantifier determination model, proved it analytically using the optimal control method and investigated the solution equivalence to the minimax problem for the RIM quantifier. However, Hong [11] recently provided a modified model for the general RIM quantifier model and the correct formulation of Liu's result.

Amin and Emrouznejad [1] have introduced the following the extended minimax disparity OWA operator model to determine the OWA operator weights:

$$\text{Minimize} \quad \max_{i\in\{1,\cdots,n-1\},\, j\in\{i+1,\cdots,n\}} |w_i - w_j|$$

$$\text{subject to} \quad orness(W) = \sum_{i=1}^{n} \frac{n-i}{n-1} w_i = \alpha, \ 0 \leq \alpha \leq 1,$$

$$w_1 + \cdots + w_n = 1, 0 \leq w_i, i = 1, \cdots, n.$$

In this paper, we propose a corresponding extended minimax disparity model for RIM quantifier determination under given orness level and prove it analytically. This paper is organized as follows: Section 2 presents the preliminaries and Section 3 reviews some models for the RIM quantifier problems and propose the extended minimax disparity model for the RIM quantifier problem. In Section 4, we prove the extended minimax disparity model problem mathematically for the case in which the generating functions are Lesbegue integrable functions.

2. Preliminaries

Yager [15] introduced a new aggregation technique based on the OWA operators. An OWA operator of dimension n is a function $F : R^n \rightarrow R$ that has an associated weighting vector $W = (w_1, \cdots, w_n)^T$ of having the properties $0 \leq w_i \leq 1$, $i = 1, \cdots, n$, $w_1 + \cdots + w_n = 1$, and such that

$$F(a_1, \cdots, a_n) = \sum_{i=1}^{n} w_i b_i,$$

where b_j is the jth largest element of the collection of the aggregated objects $\{a_1, \cdots, a_n\}$. In [15], Yager defined a measure of "orness" associated with the vector W of an OWA operator as

$$orness(W) = \sum_{i=1}^{n} \frac{n-i}{n-1} w_i,$$

and it characterizes the degree to which the aggregation is like an *or* operation.

The RIM quantifiers was introduced by Yager [16] as a method for obtaining the OWA weight vectors via fuzzy linguistic quantifiers. The RIM quantifiers can provide information aggregation procedures guided by a dimension independent description and verbally expressed concepts of the desired aggregation.

Definition 1 ([14]). *A fuzzy subset Q is called a **RIM quantifier** if $Q(0) = 0, Q(1) = 1$ and $Q(x) \geq Q(y)$ for $x > y$.*

The quantifier *for all* is represented by the fuzzy set

$$Q_*(r) = \begin{cases} 1, & x = 1, \\ 0, & x \neq 1. \end{cases}$$

The quantifier *there exist*, not none, is defined as

$$Q^*(r) = \begin{cases} 0, & x = 0, \\ 1, & x \neq 0. \end{cases}$$

Both of these are examples of RIM quantifier. To analyze the relationship between OWA and RIM quantifier, a generating function representation of RIM quantifier was proposed.

Definition 2. *For $f(t)$ on [0, 1] and a RIM quantifier $Q(x)$, $f(t)$ is called **generating function** of $Q(x)$, if it satisfies*

$$Q(x) = \int_0^x f(t)dt$$

where $f(t) \geq 0$ and $\int_0^1 f(t)dt = 1$.

If $Q(x)$ is an absolutely continuous function, then $f(x)$ is a Lesbegue integrable function; moreover, $f(x)$ is unique in the sense of "almost everywhere" in abbreviated form, *a.e.*

Yager extended the *orness* measure of OWA operator, and defined the *orness* of a RIM quantifier [16].

$$orness(Q) = \int_0^1 Q(x)dx = \int_0^1 (1-t)f(t)dt.$$

As the RIM quantifier can be seen as the continuous form of OWA operator with generating function, OWA optimization problem is extended to the RIM quantifier case.

The definitions of *essential supremum* and *essential infimum* [21] of f are as follows:

$$ess\ supf = \inf\{t : |\{x \in [0,1] : f(x) > t\}| = 0\},$$

$$ess\ inff = \sup\{t : |\{x \in [0,1] : f(x) < t\}| = 0\},$$

where $|E|$ is the Lebesgue measure of the Lebesgue measurable set E.

3. Models for the RIM Quantifier Problems

Fullér and Majlender [8] proposed the minimum variance model, which minimizes the variance of OWA operator weights under a given level of orness. Their method requires the proof of the following mathematical programming problem:

$$\text{Minimize} \quad D(W) = \frac{1}{n}\sum_{i=1}^{n-1}\left(w_i - \frac{1}{n}\right)^2$$

$$\text{subject to} \quad orness(W) = \sum_{i=1}^{n}\frac{n-i}{n-1}w_i = \alpha,\ 0 \le \alpha \le 1,$$

$$w_1 + \cdots + w_n = 1, 0 \le w_i, i = 1, \cdots, n.$$

Liu [19,24] extended the minimum variance problem for OWA operator to the RIM quantifier problem case:

$$\text{Minimize} \quad D_f = \int_0^1 f^2(r)dr - 1$$

$$\text{subject to} \quad \int_0^1 rf(r)dr = 1 - \alpha,\ 0 < \alpha < 1,$$

$$\int_0^1 f(r)dr = 1,\ f(r) \ge 0.$$

Wang and Parkan [13] proposed the minimax disparity problem as follows:

$$\text{Minimize} \quad \max_{i \in \{1, \cdots, n-1\}} |w_i - w_{i+1}|$$

$$\text{subject to} \quad orness(W) = \sum_{i=1}^{n}\frac{n-i}{n-1}w_i = \alpha,\ 0 \le \alpha \le 1,$$

$$w_1 + \cdots + w_n = 1, 0 \le w_i, i = 1, \cdots, n.$$

Similar to the minimax disparity OWA operator problem, Hong [11] proposed the minimax disparity RIM quantifier problem as follows:

$$\text{Minimize} \quad ess\ sup_{t \in [0,1]} |f'(t)|$$

$$\text{subject to} \quad \int_0^1 rf(r)dr = 1 - \alpha,\ 0 < \alpha < 1,$$

$$\int_0^1 f(r)dr = 1,\ \text{absolutely continuous } f(r) \ge 0.$$

Wang et al. [14] have introduced the following least squares deviation (LSD) method as an alternative approach to determine the OWA operator weights.

$$\text{Minimize} \quad \sum_{i=1}^{n-1} (w_i - w_{i-1})^2$$

$$\text{subject to} \quad orness(W) = \sum_{i=1}^{n} \frac{n-i}{n-1} w_i = \alpha, \ 0 \le \alpha \le 1,$$

$$w_1 + \cdots + w_n = 1, 0 \le w_i, i = 1, \cdots, n.$$

Hong [25] proposed the following corresponding least squares disparity RIM quantifier problem under a given orness level:

$$\text{Minimize} \quad D_f = \int_0^1 (f')^2(r)dr$$

$$\text{subject to} \quad \int_0^1 (1-r)f(r)dr = \alpha, \ 0 < \alpha < 1,$$

$$\int_0^1 f(r)dr = 1,$$

$$f(r) > 0.$$

Recently, Amin and Emrouznejad [1] proposed a problem of minimizing the maximum disparity of any distinct pairs of weights instead of adjacent weights. that is:

$$\text{Minimize} \quad \max_{i \in \{1, \cdots, n-1\}, \, j \in \{i+1, \cdots, n\}} |w_i - w_j| \tag{1}$$

$$\text{subject to} \quad orness(W) = \sum_{i=1}^{n} \frac{n-i}{n-1} w_i = \alpha, \ 0 \le \alpha \le 1,$$

$$w_1 + \cdots + w_n = 1, 0 \le w_i, i = 1, \cdots, n.$$

We consider the following easy important fact.
Note

$$max_{i \in \{1, \cdots, n-1\}, \, j \in \{i+1, \cdots, n\}} |w_i - w_j| = max \ w_i - min \ w_i.$$

For this, first it is trivial that

$$max_{i \in \{1, \cdots, n-1\}, \, j \in \{i+1, \cdots, n\}} |w_i - w_j| \le max \ w_i - min \ w_i.$$

Next, suppose that $max \ w_i = w_{i_0}$, $min \ w_i = w_{j_0}$. If $i_0 < j_0$, then

$$
\begin{aligned}
max \ w_i - min \ w_i \ &= \ w_{i_0} - w_{j_0} \\
&= \ |w_{i_0} - w_{j_0}| \\
&\le \ max_{i \in \{1, \cdots, n-1\}, \, j \in \{i_0+1, \cdots, n\}} |w_i - w_j|
\end{aligned}
$$

If $i_0 > j_0$, then

$$
\begin{aligned}
max \ w_i - min \ w_i \ &= \ w_{i_0} - w_{j_0} \\
&= \ |w_{j_0} - w_{i_0}| \\
&\le \ max_{i \in \{1, \cdots, n-1\}, \, j \in \{j_0+1, \cdots, n\}} |w_i - w_j|.
\end{aligned}
$$

and hence the equality holds.

Then the corresponding extended minimax disparity model for RIM quantifier problem with given orness level can be proposed as follows:

$$\text{Minimize} \quad ess\ sup\ f - ess\ inf\ f \tag{2}$$

$$\text{subject to} \quad \int_0^1 rf(r)dr = 1 - \alpha, \ \ 0 < \alpha < 1,$$

$$\int_0^1 f(r)dr = 1, \ \ f(r) \geq 0.$$

4. Relation of Solutions between OWA Operator Model and RIM Quantifier Model

The following result is the solution of the extended minimax OWA operator problem given by Hong [26].

Theorem 1 ($n = 2k$:even). *An optimal weight for the constrained optimization problem (2) for a given level of* $\alpha = orness(W)$ *should satisfy the following equation:*

$$H(\alpha) = Minimize \left\{ \max_{i \in \{1,\cdots,n-1\}, \ j \in \{i+1,\cdots,n\}} |w_i - w_j| \right\} = \left| \frac{(1 - 2\alpha)(n-1)}{(n-m)m} \right|$$

$$w_1^* = w_2^* = \cdots = w_m^*, \ w_{k+1}^* = w_{k+2}^* = \cdots = w_n^*,$$

where

$$w_1^* = \frac{m - (1 - 2\alpha)(n-1)}{nm}$$

and

$$w_{m+1}^* = \frac{n - m - (2\alpha - 1)(n-1)}{n(n-m)}.$$

Here m satisfies the following:

$$m = \begin{cases} \lceil (1 - 2\alpha)(n-1) \rceil, & if \quad 0 \leq \alpha \leq \frac{n-2}{4(n-1)}, \\ k, & if \quad \frac{n-2}{4(n-1)} \leq \alpha \leq \frac{3n-2}{4(n-1)}, \\ n - \lceil (2\alpha - 1)(n-1) \rceil, & if \quad \frac{3n-2}{4(n-1)} \leq \alpha \leq 1. \end{cases}$$

where $\lceil x \rceil = m + 1 \Longleftrightarrow m < x \leq m + 1$ *for any integer m.*

Can we get a hint about the solution of the extended minimax Rim quantifier problem? Here, we suggest an idea.

For a given associated weighting vector $W_n = (w_1, \cdots, w_n)$ of having the property $w_1 + \cdots + w_n = 1$, $0 \leq w_i \leq 1$, $i = 1, \cdots, n$, we define a generating function $f(t)$

$$f_{W_n}(x) = nw_i, \quad x \in \left[\frac{i}{n}, \frac{i+1}{n} \right), \ \ i = 0, 1, \cdots, n - 1,$$

having the property $\int_0^1 f_W^n(x)dx = 1$ and let

$$f^*(x) = \lim_{n \to \infty} = f_{W_n}(x).$$

Can this function $f^*(x)$ be a solution of the corresponding extended minimax Rim quantifier problem? Maybe, yes! Let's try to follow this idea.

For given $W_n^* = (w_1^*, \cdots, w_n^*)$ from above Theorem 1, we have for $0 < \alpha \leq \frac{1}{4}$,

$$f_{W_n^*}(x) = \begin{cases} \frac{\lceil(1-2\alpha)(n-1)\rceil - (1-2\alpha)(n-1)}{\lceil(1-2\alpha)(n-1)\rceil}, & \text{if} \quad x \in \left[0, \frac{\lceil(1-2\alpha)(n-1)\rceil}{n}\right) \\ \frac{n - \lceil(1-2\alpha)(n-1)\rceil - (2\alpha-1)(n-1)}{n - \lceil(1-2\alpha)(n-1)\rceil}, & \text{if} \quad x \in \left[\frac{\lceil(1-2\alpha)(n-1)\rceil}{n}, 1\right]. \end{cases}$$

for $\frac{1}{4} \leq \alpha \leq \frac{3}{4}$,

$$f_{W_n^*}(x) = \begin{cases} \frac{n/2 - (1-2\alpha)(n-1)}{n/2}, & \text{if} \quad x \in \left[0, \frac{1}{2}\right) \\ \frac{n/2 - (2\alpha-1)(n-1)}{(n/2)}, & \text{if} \quad x \in \left[\frac{1}{2}, 1\right]. \end{cases}$$

for $3/4 \leq \alpha \leq 1$,

$$f_{W_n^*}(x) = \begin{cases} \frac{n - \lceil(2\alpha-1)(n-1)\rceil - (1-2\alpha)(n-1)}{n - \lceil(2\alpha-1)(n-1)\rceil}, & \text{if} \quad x \in \left[0, 1 - \frac{\lceil(1-2\alpha)(n-1)\rceil}{n}\right) \\ \frac{\lceil(2\alpha-1)(n-1)\rceil - (2\alpha-1)(n-1)}{\lceil(2\alpha-1)(n-1)\rceil}, & \text{if} \quad x \in \left[1 - \frac{\lceil(1-2\alpha)(n-1)\rceil}{n}, 1\right]. \end{cases}$$

Let $\lim_{n \to \infty} f_{W_n^*}(x) = f^*(x)$, then

1. for $0 < \alpha \leq \frac{1}{4}$,

$$f^*(r) = \begin{cases} 0, & \text{if} \quad r \in [0, 1-2\alpha), \\ \frac{1}{2\alpha}, & \text{if} \quad r \in [1-2\alpha, 1]. \end{cases}$$

2. for $\frac{1}{4} \leq \alpha \leq \frac{3}{4}$,

$$f^*(r) = \begin{cases} 4\alpha - 1, & \text{if} \quad r \in \left[0, \frac{1}{2}\right), \\ 3 - 4\alpha, & \text{if} \quad r \in \left[\frac{1}{2}, 1\right]. \end{cases}$$

3. for $\frac{3}{4} < \alpha \leq 1$,

$$f^*(r) = \begin{cases} \frac{1}{2(1-\alpha)}, & \text{if} \quad r \in [0, 2\alpha], \\ 0, & \text{elsewhere.} \end{cases}$$

In the following section, we will show that f^* can be the solution of the extended minimax RIM quantifier problem.

5. Proof of the Extended Minimax RIM Quantifier Problem

In this section, we prove the following main result.

Theorem 2. *The optimal solution for problem (2) for given orness level α is the weighting function f^* such that*

1. *for $0 < \alpha \leq \frac{1}{4}$,*

$$f^*(r) = \begin{cases} 0 \ a.e., & \text{if} \quad r \in [0, 1-2\alpha), \\ \frac{1}{2\alpha} \ a.e., & \text{if} \quad r \in [1-2\alpha, 1]. \end{cases}$$

2. *for $\frac{1}{4} \leq \alpha \leq \frac{3}{4}$,*

$$f^*(r) = \begin{cases} 4\alpha - 1 \ a.e., & \text{if} \quad r \in \left[0, \frac{1}{2}\right), \\ 3 - 4\alpha \ a.e., & \text{if} \quad r \in \left[\frac{1}{2}, 1\right]. \end{cases}$$

3. for $\frac{3}{4} < \alpha \leq 1$,

$$f^*(r) = \begin{cases} \frac{1}{2(1-\alpha)} \ a.e., & if \quad r \in [0, 2\alpha], \\ 0 \ a.e., & elsewhere. \end{cases}$$

and

$$H(\alpha) = \text{Minimize} \ |ess \ sup f - ess \ inf f| = \begin{cases} \frac{1}{2\alpha} & if \quad 0 < \alpha \leq \frac{1}{4}, \\ 4|(1-2\alpha)| & if \quad \frac{1}{4} \leq \alpha \leq \frac{3}{4}, \\ \frac{1}{2\alpha} & if \quad \frac{3}{4} < \alpha \leq 1. \end{cases}$$

We need the following two lemma's to prove the main result. We denote $D_f(x) = \int_0^x f(t)dt$, $0 \leq x \leq 1$ and $E(f) = \int_0^1 rf(r)dr$.

The following result is known.

Lemma 1. $E(f) = \int_0^1 (1 - D_f(t))dt$.

Lemma 2. *Let ess inf $f = \beta_0 \geq 0$ and ess sup $f = \beta_1 > 0$ such that $\int_0^1 f(r)dr = 1$ and define a function f_0 as*

$$f_0(r) = \begin{cases} \beta_0 \ a.e., & if \quad r \in [0, c_0), \\ \beta_1 \ a.e., & if \quad r \in [c_0, 1]. \end{cases}$$

for some $c_0 \in (0,1)$ such that $\int_0^1 f_0(r)dr = 1$. Then we have $E(f) \leq E(f_0)$ and the equality holds iff $f = f_0$ a.e.

Proof. The result follows immediately from Lemma 1 if we show that $D_{f_0}(x) \leq D_f(x), x \in [0,1]$. It is clear that $D_{f_0}(x) \leq D_f(x), x \in [0, c_0]$. Suppose that there exists a point $t_0 \in (c_0, 1)$ such that $D_{f_0}(t_0) > D_f(t_0)$. Then

$$\int_{t_0}^1 \beta_1 dr = \int_{t_0}^1 f_0(r)dr = 1 - D_{f_0}(t_0) < 1 - D_f(t_0) = \int_{t_0}^1 f(r)dr$$

which implies $ess \ sup_{(t_0,1)} f > \beta_1$. It is a contradiction. $\square$

Proof of Theorem 2. If $\alpha = \frac{1}{2}$, we clearly have the optimal solution is $f^*(r) = 1$ a.e. for $r \in [0,1]$. Note that $ess \ inf f^* < 1 < ess \ sup f^*$ for $\alpha \in \left(0, \frac{1}{2}\right)$. Without loss of generality, we can assume that $\alpha \in \left(0, \frac{1}{2}\right)$, since if a weighting function $f^*(r)$ is optimal to problem (2) for some given level of preference $\alpha \in \left(0, \frac{1}{2}\right]$, then $f^*(1-r)$ is optimal to the problem (2) for a given level of preference $1 - \alpha$. Indeed, since $D_f = D_{fR}$, $\int_0^1 f(r)dr = \int_0^1 f^R(r)dr$ and $E(f^R) = 1 - E(f)$, where $f^R(r) = f(1-r)$ hence for $\alpha > \frac{1}{2}$, we can consider problem (2) for the level of preference with index $1 - \alpha$, and then take the reverse of that optimal solution. We can easily check that the weighting functions, f^*, given above are feasible for problem (2). We show that f^* is the unique optimal solution for a given α. Let nonnegative function f satisfy $1 = \int_0^1 f(r)dr$ and $E(f) = \int_0^1 rf(r)dr = 1 - \alpha$. Let $ess \ inf f = \beta_0$ and $ess \ sup f = \beta_1$.

Case (A): $\alpha \in \left(0, \frac{1}{4}\right]$.

We note that *ess inf* $f^* -$ *ess inf* $f_* = \frac{1}{2\alpha}$. We will show that $\beta_1 - \beta_0 \geq \frac{1}{2\alpha}$. To show this, we define a function f_0 as

$$f_0(r) = \begin{cases} \beta_0 & \text{if} \quad r \in [0, x_0), \\ \beta_1 & \text{if} \quad r \in [x_0, 1], \end{cases}$$

for some $x_0 \in (0, 1)$ such that $\int_0^1 f_0(r)dr = 1$. Then by Lemma 2, $E(f) \leq E(f_0)$. Suppose that $\beta_1 - \beta_0 < \frac{1}{2\alpha}$ and define another function f_0^* as

$$f_0^*(r) = \begin{cases} \beta_0 & \text{if} \quad r \in [0, x_0^*), \\ \beta_0 + \frac{1}{2\alpha} & \text{if} \quad r \in [x_0^*, 1], \end{cases}$$

for some $x_0^* \in (0, 1)$ such that $\int_0^1 f_0^*(r)dr = 1$. Then $E(f_0) < E(f_0^*)$. We note that $1 = \beta_0 x_0^* + (1 - x_0^*)(\beta_0 + \frac{1}{2\alpha})$. Then

$$x_0^* = 2\alpha\beta_0 + 1 - 2\alpha. \tag{3}$$

We know that

$$\begin{aligned} E(f_0^*) &= \beta_0 \int_0^{x_0^*} x\,dx + \left(\beta_0 + \frac{1}{2\alpha}\right) \int_{x_0^*}^1 x\,dx \\ &= \frac{\beta_0}{2} + \frac{1}{4\alpha} - \frac{x_0^{*2}}{4\alpha} \end{aligned}$$

and

$$E(f^*) = \frac{1}{2\alpha} \int_{1-2\alpha}^1 x\,dx = 1 - \alpha.$$

And we have

$$\begin{aligned} E(f^*) - E(f_0^*) &= \frac{1}{2}\frac{x_0^{*2}}{2\alpha} - \frac{1}{2}\frac{(1-2\alpha)^2}{2\alpha} - \frac{\beta_0}{2} \\ &= \frac{1}{2}\left[\frac{1}{2\alpha}x_0^{*2} - \frac{(1-2\alpha)^2}{2\alpha} - \beta_0\right] \\ &= \frac{1}{2}\left[\frac{1}{2\alpha}(2\alpha\beta_0 + 1 - 2\alpha)^2 - \frac{(1-2\alpha)^2}{2\alpha} - \beta_0\right] \\ &= \frac{\beta_0}{2}\left[2\alpha\beta_0 + 2(1 - 2\alpha) - 1\right] \\ &\geq 0 \end{aligned}$$

where the third equality comes from (3) and the last inequality comes from the facts that $1 - 2\alpha \geq \frac{1}{2}$, $\beta_0 \geq 0$ and $\alpha > 0$. This proves $E(f) < E(f_0^*) \leq E(f^*) = 1 - \alpha$, which is a contradiction. Hence f^* is an optimal solution for the case of $\alpha \in \left(0, \frac{1}{4}\right]$.

Case (B): $\alpha \in \left(\frac{1}{4}, \frac{1}{2}\right)$.

We note that *ess inf* $f^* -$ *ess inf* $f_* = 4(1 - 2\alpha)$. We will show that $\beta_1 - \beta_0 \geq 4(1 - 2\alpha)$. As in the Case (A), we define a function f_0 as

$$f_0(r) = \begin{cases} \beta_0 & \text{if} \quad r \in [0, x_0), \\ \beta_1 & \text{if} \quad r \in [x_0, 1], \end{cases}$$

for some $x_0 \in (0,1)$ such that $\int_0^1 f_0(r)dr = 1$. Then by lemma 2, $E(f) \leq E(f_0)$. Suppose that $\beta_1 - \beta_0 < \frac{1}{2\alpha}$ and define another function f_1^* as

$$f_1^*(r) = \begin{cases} \beta_0 & \text{if} \quad r \in [0, x_1^*), \\ \beta_0 + 4(1 - 2\alpha) & \text{if} \quad r \in [x_1^*, 1], \end{cases}$$

for some $x_1^* \in (0,1)$ such that $\int_0^1 f_1^*(r)dr = 1$. Then, since $x_0 < x_1^*$, by lemma 2 $E(f_0) < E(f_1^*)$. We note that $1 = \beta_0 x_1^* + (1 - x_1^*)(\beta_0 + 4(1 - 2\alpha))$. Then

$$x_1^* = 1 + \frac{\beta_0 - 1}{4(1 - 2\alpha)}$$

and

$$x_1^{*2} = 1 + \frac{\beta_0 - 1}{2(1 - 2\alpha)} + \frac{(\beta_0 - 1)^2}{16(1 - 2\alpha)^2} \tag{4}$$

We know that

$$\begin{aligned} E(f_0^*) &= \beta_0 \int_0^{x_1^*} x \, dx + (\beta_0 + 4(1 - 2\alpha)) \int_{x_1^*}^1 x \, dx \\ &= \frac{1}{2}[\beta_0 + 4(1 - 2\alpha)] - 2(1 - 2\alpha)x_1^{*2} \end{aligned}$$

and

$$\begin{aligned} E(f^*) &= (4\alpha - 1) \int_0^{\frac{1}{2}} x \, dx + (3 - 4\alpha) \int_{\frac{1}{2}}^1 x \, dx \\ &= 1 - \alpha. \end{aligned}$$

Then we have that

$$\begin{aligned} E(f^*) - E(f_1^*) &= 3\alpha - 1 - \frac{\beta_0}{2} + 2(1 - 2\alpha)x_1^{*2} \\ &= \frac{(\beta_0 - 1)^2}{8(1 - 2\alpha)} + \frac{\beta_0}{2} - \alpha \\ &= \frac{[\beta_0 - (4\alpha - 1)]^2}{8(1 - 2\alpha)} \\ &\geq 0 \end{aligned}$$

where the second equality comes from (4) and hence $E(f) < E(f_1^*) \leq E(f^*) = 1 - \alpha$, which is a contradiction. This completes the proof. $\square$

6. Conclusions

Previous studies have suggested a number of methods for obtaining optimal solution of the RIM quantifier problem. This paper proposes the extended minimax disparity RIM quantifier problem under a given orness level. We completely prove it analytically.

Funding: This research was supported by Basic Science Research Program through the National 247 Research Foundation of Korea (NRF) funded by the Ministry of Education (2017R1D1A1B03027869).

Conflicts of Interest: The authors declare no conflict of interest.

References

1. Amin, G.R.; Emrouznejad, A. An extended minimax disparity to determine the OWA operator weights. *Comput. Ind. Eng.* **2006**, *50*, 312–316. [CrossRef]
2. Amin, G.R. Notes on priperties of the OWA weights determination model. *Comput. Ind. Eng.* **2007**, *52*, 533–538. [CrossRef]
3. Emrouznejad, A.; Amin, G.R. Improving minimax disparity model to determine the OWA operator weights. *Inf. Sci.* **2010**, *180*, 1477–1485. [CrossRef]
4. Emrouznejad, A. MP-OWA: The most preferred OWA operator. *Knowl. Based Syst.* **2008**, *21*, 847–851. [CrossRef]
5. Emrouznejad, A.; Marra, M. Ordered Weighted Averaging Operators 1988-2014: A citation-based literature survey. *Int. J. Intell. Syst.* **2014**, *29*, 994–1014. [CrossRef]
6. Filev, D.; Yager, R.R. On the issue of obtaining OWA operator weights. *Fuzzy Sets Syst.* **1988**, *94*, 157–169. [CrossRef]
7. Fullér, R.; Majlender, P. An analytic approach for obtaining maximal entropy OWA operators weights. *Fuzzy Sets Syst.* **2001**, *124*, 53–57. [CrossRef]
8. Fullér, R.; Majlender, P. On obtaining minimal variability OWA operator weights. *Fuzzy Sets Syst.* **2003**, *136*, 203–215. [CrossRef]
9. O'Hagan, M. Aggregating template or rule antecedents in real-time expert systems with fuzzy set logic. In Proceedings of the 22nd annual IEEE Asilomar Conf. on Signals, Systems, Computers, Pacific Grove, CA, USA, 31 October–2 November 1988; pp. 681–689.
10. Hong, D.H. A note on the minimal variability OWA operator weights. *Int. J. Uncertainty, Fuzziness Knowl. Based Syst.* **2006**, *14*, 747–752. [CrossRef]
11. Hong, D.H. A note on solution equivalence to general models for RIM quantifier problems. *Fuzzy Sets Syst.* **2018**, *332*, 25–28. [CrossRef]
12. Wheeden, R.L.; Zygmund, A. *Measure and Integral: An Introduction to Real Analysis*; Marcel Dekker, Inc.: New York, NY, USA, 1977.
13. Wang, Y.M.; Parkan, C. A minimax disparity approach obtaining OWA operator weights. *Inf. Sci.* **2005**, *175*, 20–29. [CrossRef]
14. Wang, Y.M.; Luo, Y.; Liu, X. Two new models for determining OWA operater weights. *Comput. Ind. Eng.* **2007**, *52*, 203–209. [CrossRef]
15. Yager, R.R. Ordered weighted averaging aggregation operators in multi-criteria decision making. *IEEE Trans. Syst. Man Cybern.* **1988**, *18*, 183–190. [CrossRef]
16. Yager, R.R. OWA aggregation over a continuous interval argument with application to decision making. *IEEE Trans. Syst. Man Cybern. Part B* **2004**, *34*, 1952–1963. [CrossRef]
17. Yager, R.R. Families of OWA operators. *Fuzzy Sets Syst.* **1993**, *59*, 125–148. [CrossRef]
18. Yager, R.R.; Filev, D. Induced ordered weighted averaging operators. *IEEE Trans. Syst. Man Cybern. Part B Cybern.* **1999**, *29*, 141–150. [CrossRef]
19. Liu, X. On the maximum entropy parameterized interval approximation of fuzzy numbers. *Fuzzy Sets Syst.* **2006**, *157*, 869–878. [CrossRef]
20. Liu, X.; Da, Q. On the properties of regular increasing monotone (RIM) quantifiers with maximum entropy. *Int. J. Gen. Syst.* **2008**, *37*, 167–179. [CrossRef]
21. Liu, X.; Lou, H. On the equivalence of some approaches to the OWA operator and RIM quantifier determination. *Fuzzy Sets Syst.* **2007**, *159*, 1673–1688. [CrossRef]
22. Hong, D.H. The relationship between the minimum variance and minimax disparity RIM quantifier problems. *Fuzzy Sets Syst.* **2011**, *181*, 50–57. [CrossRef]
23. Hong, D.H. The relationship between the maximum entropy and minimax ratio RIM quantifier problems. *Fuzzy Sets Syst.* **2012**, *202*, 110–117. [CrossRef]
24. Liu, X. A general model of parameterized OWA aggregation with given orness level. *Int. J. Approx. Reason.* **2008**, *48*, 598–627. [CrossRef]

25. Hong, D.H. The general model for least square disparity RIM quantifier problems. *Fuzzy Optim. Decis. Mak.* **2019**, submitted.
26. Hong, D.H. On proving the extended minimax disparity OWA problem. *Fuzzy Sets Syst.* **2011**, *168*, 35–46. [CrossRef]

Article

The Solution Equivalence to General Models for the RIM Quantifier Problem

Dug Hun Hong

Department of Mathematics, Myongji University, Yongin 449-728, Kyunggido, Korea; dhhong@mju.ac.kr

Received: 3 March 2019; Accepted: 28 March 2019; Published: 1 April 2019

Abstract: Hong investigated the relationship between the minimax disparity minimum variance regular increasing monotone (RIM) quantifier problems. He also proved the equivalence of their solutions to minimum variance and minimax disparity RIM quantifier problems. Hong investigated the relationship between the minimax ratio and maximum entropy RIM quantifier problems and proved the equivalence of their solutions to the maximum entropy and minimax ratio RIM quantifier problems. Liu proposed a general RIM quantifier determination model and proved it analytically by using the optimal control technique. He also gave the equivalence of solutions to the minimax problem for the RIM quantifier. Recently, Hong proposed a modified model for the general minimax RIM quantifier problem and provided correct formulation of the result of Liu. Thus, we examine the general minimum model for the RIM quantifier problem when the generating functions are Lebesgue integrable under the more general assumption of the RIM quantifier operator. We also provide a solution equivalent relationship between the general maximum model and the general minimax model for RIM quantifier problems, which is the corrected and generalized version of the equivalence of solutions to the general maximum model and the general minimax model for RIM quantifier problems of Liu's result.

Keywords: OWA operator; RIM quantifier; maximum entropy; minimax ratio; generating function; minimal variability; minimax disparity; solution equivalence

1. Introduction

One of the important topics in the theory of ordered weighted averaging (OWA) operators is the determination of the associated weights. Several authors have suggested a number of methods for obtaining associated weights in various areas such as decision-making, approximate reasoning, expert systems, data mining, fuzzy systems and control [1–22]. Yager [12] proposed RIM quantifiers as a method for finding OWA weight vectors through fuzzy linguistic quantifiers. Liu [15] and Liu and Da [16] gave solutions to the maximum-entropy RIM quantifier model when the generating functions are differentiable. Liu and Lou [9] studied the equivalence of solutions to the minimax ratio and maximum-entropy RIM quantifier models, and the equivalence of solutions to the minimax disparity and minimum-variance RIM quantifier problems. Hong [17,18] gave the proof of the minimax ratio RIM quantifier problem and the minimax disparity RIM quantifier model when the generating functions are absolutely continuous. He also gave solutions to the maximum-entropy RIM quantifier model and the minimum-variance RIM quantifier model when the generating functions are Lebesgue integrable.

Based on these results, Hong [17,18] provided a relationship between the minimax disparity and minimum-variance RIM quantifier problems. He also provided a correct relationship between the minimax ratio and maximum-entropy RIM quantifier models. Liu [19] suggested a general RIM quantifier determination model and proved it analytically using the optimal control methods. He also studied the solution equivalence to the minimax problem for the RIM quantifier.

This paper investigates the general minimax model for the RIM quantifier problem for the case in which the generating functions are absolutely continuous and a generalized solution to the general minimum model for the RIM quantifier problem for the case in which the generating functions are Lebesgue integrable. Moreover, this paper provides a solution equivalent relationship between the general maximum model and the general minimax model for RIM quantifier problems and generalizes the results of Hong [17,18]. In this paper, we improve and extend Liu's theorems to be suitable for absolutely continuous generating functions. We have corrected and improved Theorem 13 [19] by using the absolutely continuous condition of generating functions and the absolute continuity condition of F' for the general minimax model for the RIM quantifier problem. Theorem 9 [19] has been improved using the Lebesgue integrability condition of generating functions and the continuity condition of F' for the general maximum model for the RIM quantifier problem.

Based on these results, we give a correct relationship between the general minimum model and the general minimax model for RIM quantifier problems.

2. Preliminaries

Yager [11] proposed a new aggregation technique based on OWA operators. An OWA operator of dimension n is a mapping $F : R^n \to R$ that has an associated weight vector $W = (w_1, \cdots, w_n)^T$ with the properties $w_1 + \cdots + w_n = 1$, $0 \le w_i \le 1$, $i = 1, \cdots, n$, such that

$$F(a_1, \cdots, a_n) = \sum_{i=1}^{n} w_i b_i,$$

where b_j is the jth largest element of the collection of the aggregated objects $\{a_1, \cdots, a_n\}$. In [11], Yager introduced a measure of "orness" associated with the weight vector W of an OWA operator:

$$orness(W) = \sum_{i=1}^{n} \frac{n-i}{n-1} w_i.$$

This measure characterizes the degree to which the aggregation is like an OR operation.

Here, *min, max,* and *average* correspond to W^*, W_* and W_A respectively, where $W^* = (1,0,\cdots,0), W_* = (0,0,\cdots,1)$ and $W_A = (1/n, 1/n, \cdots, 1/n)$. Clearly, $orness(W^*) = 1, orness(W_*) = 0$ and $orness(W_A) = 1/2$.

Yager [12] introduced RIM quantifiers as a method for obtaining OWA weight vectors through fuzzy linguistic quantifiers.

Definition 1 ([12]). *A fuzzy subset Q on the real line is called a RIM quantifier if $Q(0) = 0, Q(1) = 1$ and $Q(x) \ge Q(y)$ for $x > y$.*

The quantifier *for all* is represented by the fuzzy set

$$Q_*(r) = \begin{cases} 1, & \text{if } x = 1, \\ 0, & \text{if } x \ne 1. \end{cases}$$

The quantifier *there exists* is defined as

$$Q^*(r) = \begin{cases} 0, & \text{if } x = 0, \\ 1, & \text{if } x \ne 0. \end{cases}$$

Both of these are examples of the RIM quantifier. A generating function representation of RIM quantifiers has been proposed for analyzing the relationship between OWA operators and RIM quantifiers.

Definition 2. *For $f(t)$ on [0, 1] and the RIM quantifier $Q(x)$, $f(t)$ is called the generating function of $Q(x)$, if it satisfies*

$$Q(x) = \int_0^x f(t)dt,$$

where $f(t) \geq 0$ and $\int_0^1 f(t)dt = 1$.

If the RIM quantifier $Q(x)$ is smooth, then $f(x)$ should be continuous; if $Q(x)$ is a piecewise linear function, then $f(x)$ is a jump piecewise function of some constants; and if $Q(x)$ is an absolutely continuous function, then $f(x)$ is a Lesbegue integrable function and unique in the sense of being "almost everywhere" [23].

Yager extended the *orness* measure of OWA operators, and defined the *orness* of RIM quantifiers [10] as:

$$orness(Q) = \int_0^1 Q(x)dx = \int_0^1 (1-t)f(t)dt.$$

We see that Q_* leads to the weight vector W_*, Q^* leads to the weight vector W^*, and the ordinary *average* RIM quantifier $Q_A(x) = x$ leads to the weight vector W_A. We also have $orness(Q^*) = 1$, $rness(Q_*) = 0$, and $orness(Q_A) = 1/2$.

As the RIM quantifier can be seen as a continuous form of OWA, an operator with a generating function, the OWA optimization problem can be extended to the case of the RIM quantifier.

3. The General Model for the Minimax RIM Quantifier Problem

In this section, we consider the general model for the minimax RIM quantifier problem and generalize some results of Hong [17,18]. Hong [7] provided a modified model for the minimax RIM quantifier problem and the correct formulation of a result of Liu [19]. We summarize briefly.

* **The minimax disparity RIM quantifier problem [15,17].**

The minimax disparity RIM quantifier problem with a given orness level $0 < \alpha < 1$ consists of finding a solution $f : [0,1] \to [0,1]$ to the following optimization problem:

$$\text{Minimize} \quad \max_{t \in (0,1)} |f'(t)|,$$

$$\text{subject to} \quad \int_0^1 (1-r)f(r)dr = \alpha, \quad 0 < \alpha < 1,$$

$$\int_0^1 f(r)dr = 1,$$

$$f(r) \geq 0.$$

* **The minimax ratio RIM quantifier problem [9,18].**

The minimax ratio RIM quantifier problem with a given orness level $0 < \alpha < 1$ consists of finding a solution $f : [0,1] \to [0,1]$ to the following optimization problem:

$$\text{Minimize} \quad \max_{t \in (0,1)} \left| \frac{f'(t)}{f(t)} \right|,$$

$$\text{subject to} \quad \int_0^1 (1-r)f(r)dr = \alpha, \quad 0 < \alpha < 1,$$

$$\int_0^1 f(r)dr = 1,$$

$$f(r) > 0.$$

In regard to the above optimization problem, Liu [19] considered a general model for the minimax RIM quantifier problem:

$$\text{Minimize} \quad M_f = \max_{r \in (0,1)} \left| F''(f(r))f'(r) \right|,$$

$$\text{subject to} \quad \int_0^1 rf(r)dr = \alpha, \ 0 < \alpha < 1, \tag{1}$$

$$\int_0^1 f(r)dr = 1,$$

$$f(r) > 0,$$

where the generating functions are continuous and F is a strictly convex function on $[0, \infty)$, which is differentiable to at least the 2nd order.

The above two cases are special cases of this model with $F(x) = x^2$ and $F(x) = x \ln x$. Hong [7] gave a corrected and modified general model for the minimax RIM quantifier problem as follows:

∗ The general model for the minimax RIM quantifier problem.

$$\text{Minimize} \quad M_f = ess \, sup_{r \in (0,1)} \left| F''(f(x))f'(x) \right|,$$

$$\text{subject to} \quad \int_0^1 rf(r)dr = \alpha, \ 0 < \alpha < 1, \tag{2}$$

$$\int_0^1 f(r)dr = 1,$$

$$f(r) > 0.$$

Theorem 1. *Supposing that the generating functions are absolutely continuous, F is a strictly convex function on $[0, \infty)$, and F' is absolutely continuous, then there is a unique optimal solution for problem (2), and that the optimal solution has the form*

$$f^*(r) = \max \left\{ (F')^{-1} \left(a^*r + b^* \right), 0 \right\},$$

where a^* and b^* are determined by the constraints:

$$\begin{cases} \int_0^1 rf^*(r)dr = \alpha, \\ \int_0^1 f^*(r)dr = 1, \\ f^*(r) \geq 0. \end{cases}$$

The next example shows that the condition of F' being *absolutely continuous* on $[0, \infty)$ in Theorem 1 is essential.

Example 1. *Letting $F_1(x) = \int_0^x (C(r) + r)dr$ where $C(x)$ is a Cantor function, then $F'(x) = C(x) + x$ and $F_1''(x) = 1$ a.e. but $F_1'(x) \neq \int_0^x F_1''(r)dr$, that is, F_1' is not absolutely continuous on $[0, \infty)$. Let $F_2(x) = (1/2)x^2$, then $F_2''(x) = 1$. Since*

$$ess \, sup_{r \in (0,1)} \left| F_1''(f(x))f'(x) \right| = ess \, sup_{r \in (0,1)} \left| f'(x) \right| = ess \, sup_{r \in (0,1)} \left| F_2''(f(x))f'(x) \right|,$$

the optimal solution of problem (2) with respect to F_1 and F_1 are the same. However, since $F_1'(x) \neq F_2'(x)$, the optimal solution of problem (2) with respect to F_1 and F_1 cannot be the same by Theorem 2, which is a contradiction. This example shows the Theorem 2 is incorrect if F' is not absolutely continuous on $[0, \infty)$.

4. The General Model for the Minimum RIM Quantifier Problem

In this section, we consider the general model for the minimum RIM quantifier problem. We improve the results of Liu [19] and generalize Theorem 4 of Hong [17] and Theorem 5 of Hong [18]. Liu [19] obtained solutions to the general minimum RIM quantifier problem for the case in which the generating functions are continuous and F is differentiable to at least the 2nd order by considering a variational optimization problem using the Lagrangian multiplier method ([24], Chapter 2). In this section, we consider a generalized result for this problem.

∗ **The minimum variance RIM quantifier problem [17,18].**

The minimum variance RIM quantifier problem under a given orness level is

$$\text{Minimize} \quad D_f = \int_0^1 f^2(r)dr,$$

$$\text{subject to} \quad \int_0^1 rf(r)dr = \alpha, \ 0 < \alpha < 1,$$

$$\int_0^1 f(r)dr = 1,$$

$$f(r) > 0.$$

∗ **The maximum entropy RIM quantifier problem [9,18].**

The maximum entropy RIM quantifier problem with a given orness level $0 < \alpha < 1$ consists of finding a solution $f : [0,1] \to [0,1]$ to the following optimization problem:

$$\text{Maximize} \quad -\int_0^1 f(r)\ln f(r)dr,$$

$$\text{subject to} \quad \int_0^1 rf(r)dr = \alpha, \ 0 < \alpha < 1,$$

$$\int_0^1 f(r)dr = 1,$$

$$f(r) > 0.$$

Recently, Liu [19] considered the general model for the minimum variance and maximum entropy RIM quantifier problems, under a given orness level formulated as follows:

∗ **The general model for the minimum RIM quantifier problem.**

$$\text{Minimize} \quad V_f = \int_0^1 F(f(r))dr,$$

$$\text{subject to} \quad \int_0^1 rf(r)dr = \alpha, \ 0 < \alpha < 1, \tag{3}$$

$$\int_0^1 f(r)dr = 1,$$

$$f(r) > 0,$$

where F is a strictly convex function on $[0, \infty)$, and differentiable to at least the 2nd order.

The above two cases are special cases of the model where $F(x) = x^2$ and $F(x) = x\ln x$.

Liu (Theorem 9, [19]) proved the following problem for the case in which generating functions are continuous and F is differentiable to at least the 2nd order:

Theorem 2 (Theorem 9, [19])**.** *There is a unique optimal solution for (3), and the optimal solution has the form*

$$f^*(r) = \begin{cases} (F')^{-1}(a^*r + b^*), & \text{if } (F')^{-1}(a^*r + b^*) \geq 0, \\ 0, & \text{elsewhere}, \end{cases}$$

where a^*, b^* are determined by the constraints:

$$\begin{cases} \int_0^1 rf^*(r)dr = \alpha, \\ \int_0^1 f^*(r)dr = 1, \\ f^*(r) \geq 0. \end{cases}$$

Here, we consider a generalized result for Theorem 3 when $f(x)$ is Lebesgue integrable and F' is continuous.

Theorem 3. *Suppose that the generating functions are Lebesgue integrable, F is a strictly convex function on $[0, \infty)$, and F' is continuous. Then, there is a unique optimal solution for problem (3), and that optimal solution has the form*

$$f^*(r) = \begin{cases} (F')^{-1}(a^*r + b^*) \ \text{a.e.}, & \text{if } (F')^{-1}(a^*r + b^*) > 0, \\ 0 \ \text{a.e.}, & \text{elsewhere}, \end{cases}$$

where a^* and b^* are determined by the constraints:

$$\begin{cases} \int_0^1 rf^*(r)dr = \alpha, \\ \int_0^1 f^*(r)dr = 1, \\ f^*(r) \geq 0. \end{cases}$$

Proof. As shown in Theorem 2, we consider the case where $\alpha \in (0, 1/2]$ and assume that $\{r < 1 : f^*(r) > 0\} = [0, t)$ for some $t \in (0, 1)$ and $\{r < 1 : f^*(r) = 0\} = [t, 1)$. We also note that for $r \in [0, t]$,

$$F'(f^*(r)) = a^*r + b^*$$

and for $r \in (t, 1)$,

$$a^*r + b^* < F'(0)$$

if $F'(0)$ exists. Let the nonnegative function f satisfy $1 = \int_0^1 f(r)dr$ and $\int_0^1 rf(r)dr = \alpha$. We set $f(r) = f^*(r) + g(r)$, $r \in [0, 1]$. Then, noting that $f(r) = g(r)$, $r \in [t, 1]$, we have

$$\int_0^t g(r)dr + \int_t^1 f(r)dr = \int_0^1 g(r)dr = 0, \tag{4}$$

since $1 = \int_0^1 f(r)dr = \int_0^1 f^*(r)dr + \int_0^1 g(r)dr = 1 + \int_0^1 g(r)dr$. We also have

$$\int_0^t rg(r)dr + \int_t^1 rf(r)dr = \int_0^1 rg(r)dr = 0, \tag{5}$$

since $\alpha = \int_0^1 rf(r)dr = \int_0^1 rf^*(r)dr + \int_0^1 rg(r)dr = \alpha + \int_0^1 rg(r)dr$. We now show that

$$\int_0^1 F(f(r))\, dr \geq \int_0^1 F(f^*(r))\, dr.$$

Since $F(x) - F(x_0) \geq F'(x_0)(x - x_0)$ (the equality holds if and only if $x = x_0$), we have that

$$
\begin{aligned}
& \int_0^1 F(f(r))dr - \int_0^1 F(f^*(r))dr \\
= \; & \int_0^1 F((f^*(r) + g(r)))dr - \int_0^1 F(f^*(r))dr \\
\geq \; & \int_0^1 F'(f^*(r))g(r)dr \\
= \; & \int_0^t (a^*r + b^*)g(r)dr + \int_t^1 F'(0)g(r)dr \\
= \; & a^* \int_0^t rg(r)dr + b^* \int_0^t g(r)dr + \int_t^1 F'(0)g(r)dr \\
= \; & a^*\left(-\int_t^1 rf(r)dr\right) + b^*\left(-\int_t^1 f(r)dr\right) + \int_t^1 F'(0)g(r)dr \\
= \; & \int_t^1 (F'(0) - a^*r - b^*)f(r)dr \\
\geq \; & 0,
\end{aligned}
$$

where the fourth equality comes from (4) and (5) and the second inequality comes from the fact that $a^*r + b^* \leq F'(0)$ a.e. for $r \in [t, 1]$. The equalities hold if and only if $f^* = f$ a.e. This completes the proof. $\square$

Combining Theorems 2 and 4, we now have a solution equivalent relationship between the general minimum RIM quantifier problem and the general minimax RIM quantifier problem. This result generalizes Theorem 6 of Hong [17] and Theorem 5 of Hong [18] and provides a corrected version of Theorem 13 [19].

Theorem 4. *Suppose that the generating functions are absolutely continuous and F' is increasing and absolutely continuous. Then, the general minimum RIM quantifier problem has the same solution as the general minimax RIM quantifier problem.*

5. Numerical Example

We consider a RIM quantifier operator F which is not differentiable to at least the second order, but F' is absolutely continuous, and find an optimal solution of two RIM quantifier problems.

Let a RIM quantifier operator F be

$$
F(x) = \begin{cases} \frac{x^2}{2}, & \text{if } 0 \leq x < \frac{1}{2}, \\ x^2 - \frac{1}{2}x + \frac{1}{8}, & \text{if } \frac{1}{2} \leq x \leq 1. \end{cases}
$$

Then,

$$
F'(x) = \begin{cases} x, & \text{if } 0 \leq x < \frac{1}{2}, \\ 2x - \frac{1}{2}, & \text{if } \frac{1}{2} \leq x \leq 1. \end{cases}
$$

Hence, $F(x)$ is strictly convex and $F'(x)$ is absolutely continuous, but $F(x)$ is not the second order differentiable. Let

$$
f^*(r) = \begin{cases} (F')^{-1}(a^*r + b^*), & \text{if } (F')^{-1}(a^*r + b^*) > 0, \\ 0, & \text{elsewhere}, \end{cases}
$$

where a^* and b^* are determined by the constraints:

$$\begin{cases} \int_0^1 r f^*(r)\,dr = \alpha, \\ \int_0^1 f^*(r)\,dr = 1, \\ f^*(r) \geq 0. \end{cases} \tag{6}$$

We consider the case for $0 < \alpha \leq 1/2$. Then, $a^* \leq 0$ and $b^* > 0$.

Case (1) (See Figure 1) There exists $m, d \in [0,1]$ such that $m < d$ and

$$f^*(r) = \begin{cases} \frac{1}{2}(a^*r + b^*) + \frac{1}{4}, & \text{if } 0 \leq r \leq m, \\ a^*r + b^*, & m < r \leq d, \\ 0, & d < r \leq 1. \end{cases}$$

Since $a^*m + b^* = \frac{1}{2}$ and $a^*d + b^* = 0$, $b^* = -a^*m + \frac{1}{2}$ and $d = m - \frac{1}{2a^*}$. Hence,

$$f^*(r) = \begin{cases} \frac{1}{2}a^*(r - m) + \frac{1}{2}, & \text{if } 0 \leq r \leq m, \\ a^*(r - m) + \frac{1}{2}, & m < r \leq m - \frac{1}{2a^*}, \\ 0, & m - \frac{1}{2a^*} < r \leq 1. \end{cases}$$

From (6),

$$a^* = \frac{2m - 4 - \sqrt{2m^2 - 16m + 16}}{2m^2},$$

$$\alpha = -\frac{4m^3 a^{*3} - 12m^2 a^{*2} + 6ma^* - 1}{48a^{*2}}$$

hold. In addition, since $a^* < 0$ and $f^*(1) < 0$,

$$0 < m < 4 - \sqrt{10}, \quad 0 < \alpha < \frac{17 - 4\sqrt{10}}{12}.$$

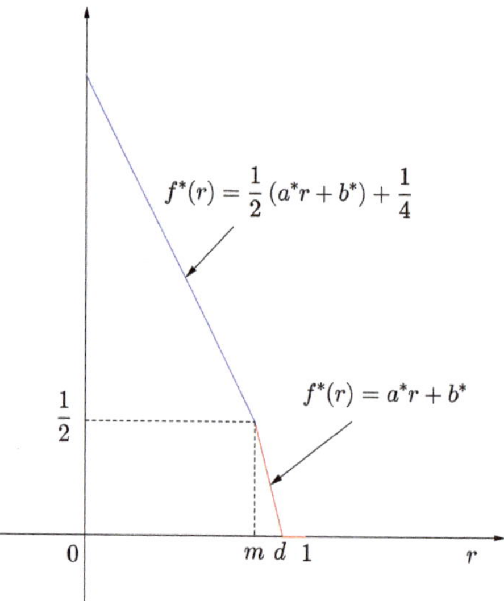

Figure 1. The graph of f^* ($0 < \alpha < \frac{17 - 4\sqrt{10}}{12}$).

Case (2) (See Figure 2) There exists $m \in [0, 1]$ such that

$$f^*(r) = \begin{cases} \frac{1}{2}(a^*r + b^*) + \frac{1}{4}, & \text{if } 0 \le r \le m, \\ a^*r + b^*, & m < r \le 1. \end{cases}$$

Since $a^*m + b^* = \frac{1}{2}$,

$$f^*(r) = \begin{cases} \frac{1}{2}a^*(r - m) + \frac{1}{2}, & \text{if } 0 \le r \le m, \\ a^*(r - m) + \frac{1}{2}, & m < r \le 1. \end{cases}$$

From (6),

$$a^* = \frac{2}{m^2 - 4m + 2},$$
$$\alpha = \frac{2m^3 + 3m^2 - 24m + 14}{12(m^2 - 4m + 2)}$$

hold. In addition, since $a^* < 0$ and $f^*(1) \ge 0$,

$$4 - \sqrt{10} \le m \le 1, \qquad \frac{17 - 4\sqrt{10}}{12} \le \alpha \le \frac{5}{12}.$$

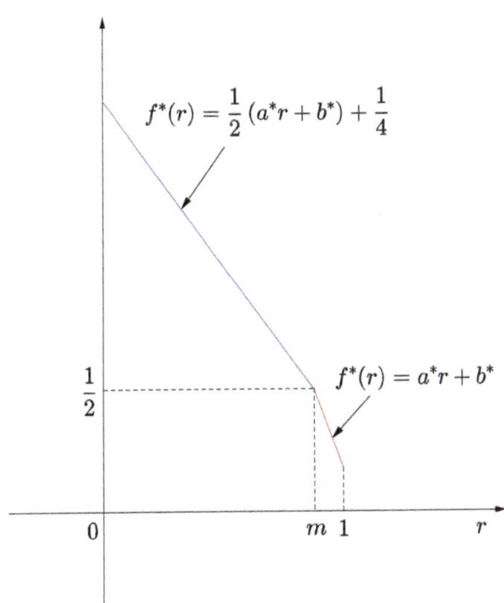

Figure 2. The graph f^* ($\frac{17 - 4\sqrt{10}}{12} \le \alpha \le \frac{5}{12}$).

Case (3) (See Figure 3) For all $0 \le r \le 1$,

$$f^*(r) = \frac{1}{2}(a^*r + b^*) + \frac{1}{4}.$$

From (6),

$$a^* = -12 + 24\alpha,$$
$$b^* = \frac{15}{2} - 12\alpha$$

hold. In addition, since $a^* \leq 0$ and $f^*(1) > \frac{1}{2}$, $\frac{5}{12} < \alpha \leq \frac{1}{2}$.

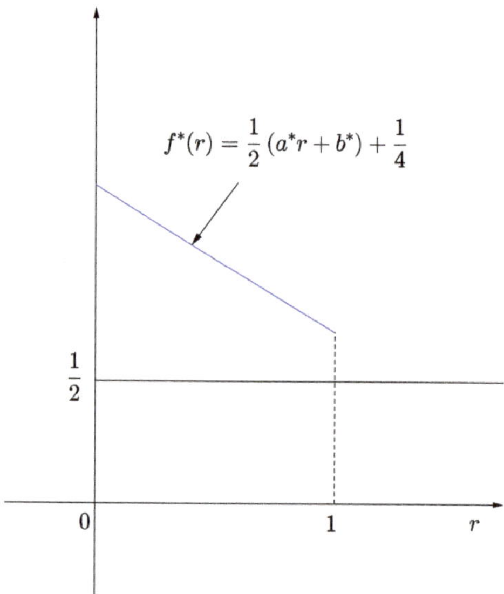

Figure 3. The graph f^* ($\frac{5}{12} < \alpha \leq \frac{1}{2}$).

6. Conclusions

In this paper, we examined the general minimax model for the RIM quantifier problem for the case in which the generating functions are absolutely continuous and a generalized solution to the general minimum model for the RIM quantifier problem for the case in which the generating functions are Lebesgue integrable. In addition, we provided a solution equivalent relationship between the general maximum model and the general minimax model for RIM quantifier problems and generalizes results of Hong based on these results. We also corrected Liu's theorems from a mathematical perspective as their theorems are not suitable for absolutely continuous generating functions.

Funding: This research was funded by the Basic Science Research Program through the National Research Foundation of Korea (NRF) grant number 2017R1D1A1B03027869.

Conflicts of Interest: The author declares no conflict of interest.

References

1. Amin, G.R.; Emrouznejad, A. An extended minimax disparity to determine the OWA operator weights. *Comput. Ind. Eng.* **2006**, *50*, 312–316. [CrossRef]
2. Amin, G.R. Notes on priperties of the OWA weights determination model. *Comput. Ind. Eng.* **2007**, *52*, 533–538. [CrossRef]
3. Emrouznejad, A.; Amin, G.R. Improving minimax disparity model to determine the OWA operator weights. *Inf. Sci.* **2010**, *180*, 1477–1485. [CrossRef]
4. Filev, D.; Yager, R.R. On the issue of obtaining OWA operator weights. *Fuzzy Sets Syst.* **1988**, *94*, 157–169. [CrossRef]
5. Fullér, R.; Majlender, P. An analytic approach for obtaining maximal entropy OWA operators weights. *Fuzzy Sets Syst.* **2001**, *124*, 53–57. [CrossRef]
6. Hagan, M.O. Aggregating template or rule antecedents in real-time expert systems with fuzzy set logic. In Proceedings of the 22nd Annual IEEE Asilomar Conference on Signals, Systems, Computers, Pacific Grove, CA, USA, 31 October–2 November 1988; pp. 681–689.

7. Hong, D.H. A note on solution equivalence to general models for RIM quantifier problems. *Fuzzy Sets Syst.* **2018**, *332*, 25–28. [CrossRef]
8. Hong, D.H. On proving the extended minimax disparity OWA problem. *Fuzzy Sets Syst.* **2011**, *168*, 35–46. [CrossRef]
9. Liu, X.; Lou, H. On the equivalence of some approaches to the OWA operator and RIM quantifier determination. *Fuzzy Sets Syst.* **2007**, *159*, 1673–1688. [CrossRef]
10. Wang, Y.M.; Parkan, C. A minimax disparity approach obtaining OWA operator weights. *Inf. Sci.* **2005**, *175*, 20–29. [CrossRef]
11. Yager, R.R. Ordered weighted averaging aggregation operators in multi-criteria decision making. *IEEE Trans. Syst. Man Cybern.* **1988**, *18*, 183–190. [CrossRef]
12. Yager, R.R. OWA aggregation over a continuous interval argument with application to decision making. *IEEE Trans. Syst. Man Cybern. Part B* **2004**, *34*, 1952–1963. [CrossRef]
13. Yager, R.R. Families of OWA operators. *Fuzzy Sets Syst.* **1993**, *59*, 125–148. [CrossRef]
14. Yager, R.R.; Filev, D. Induced ordered weighted averaging operators. *IEEE Trans. Syst. Man Cybern. Part B* **1999**, *29*, 141–150. [CrossRef] [PubMed]
15. Liu, X. On the maximum entropy parameterized interval approximation of fuzzy numbers. *Fuzzy Sets Syst.* **2006**, *157*, 869–878. [CrossRef]
16. Liu, X.; Da, Q. On the properties of regular increasing monotone (RIM) quantifiers with maximum entropy. *Int. J. Gen. Syst.* **2008**, *37*, 167–179. [CrossRef]
17. Hong, D.H. The relationship between the minimum variance and minimax disparity RIM quantifier problems. *Fuzzy Sets Syst.* **2011**, *181*, 50–57. [CrossRef]
18. Hong, D.H. The relationship between the maximum entropy and minimax ratio RIM quantifier problems. *Fuzzy Sets Syst.* **2012**, *202*, 110–117. [CrossRef]
19. Liu, X. A general model of parameterized OWA aggregation with given orness level. *Int. J. Approx. Reason.* **2008**, *48*, 598–627. [CrossRef]
20. Fullér, R.; Majlender, P. On obtaining minimal variability OWA operator weights. *Fuzzy Sets Syst.* **2003**, *136*, 203–215. [CrossRef]
21. Sang, X.; Liu, X. An analytic approach to obtain the least square deviation OWA operater weights. *Fuzzy Sets Syst.* **2014**, *240*, 103–116. [CrossRef]
22. Wang, Y.M.; Luo, Y.; Liu, X. Two new models for determining OWA operater weights. *Comput. Ind. Eng.* **2007**, *52*, 203–209. [CrossRef]
23. Wheeden, R.L.; Zygmund, A. *Measure and Integral: An Introduction to Real Analysis*; Marcel Dekker, Inc.: New York, NY, USA, 1977.
24. Rustagi, J.S. *Variational Methods in Statistics*; Academic Press: New York, NY, USA, 1976.

 *symmetry*

Correction

Correction: Kim, T.; Khan, W.A.; Sharma, S.K.; Ghayasuddin, M. A Note on Parametric Kinds of the Degenerate Poly-Bernoulli and Poly-Genocchi Polynomials. Symmetry 2020, *12*(4), 614

Taekyun Kim [1], Waseem A. Khan [2], Sunil Kumar Sharma [3,*] and Mohd Ghayasuddin [4]

[1] Department of Mathematics, Kwangwoon University, Seoul 139-701, Korea; tkkim@kw.ac.kr
[2] Department of Mathematics and Natural Sciences, Prince Mohammad Bin Fahd University, P.O Box 1664, Al Khobar 31952, Saudi Arabia; wkhan1@pmu.edu.sa
[3] College of Computer and Information Sciences, Majmaah University, Majmaah 11952, Saudi Arabial
[4] Department of Mathematics, Integral University Campus, Shahjahanpur 242001, India; ghayas.maths@gmail.com
* Correspondence: s.sharma@mu.edu.sa

Received: 24 May 2020; Accepted: 25 May 2020; Published: 26 May 2020

(This article belongs to the Special Issue Current Trends in Symmetric Polynomials with Their Applications II)

The authors wish to provide the following information: on page 15, add funding information [1].

Funding: The authors extend their appreciation to the Deanship of Scientific Research at Majmaah University for funding this work under project number (RGP-2019-25).

The authors would like to apologize for any inconvenience caused to the readers by these changes. The changes do not affect the scientific results. The manuscript will be updated and the original will remain online on the article webpage, with a reference to this Correction.

Reference

1. Kim, T.; Khan, W.A.; Sharma, S.K.; Ghayasuddin, M. A Note on Parametric Kinds of the Degenerate Poly-Bernoulli and Poly-Genocchi Polynomials. *Symmetry* **2020**, *12*, 614. [CrossRef]

MDPI

St. Alban-Anlage 66

4052 Basel

Switzerland

Tel. +41 61 683 77 34

Fax +41 61 302 89 18

www.mdpi.com

Symmetry Editorial Office

E-mail: symmetry@mdpi.com

www.mdpi.com/journal/symmetry